Ordering Information: See page 142 for information on different versions of this book and
ordering, on library purchases, and on quantity discounts for Alexander Technique teachers.

Library and Archives Canada Cataloguing in Publication

Gorman, David, 1950-
    Looking at ourselves : exploring and evolving in the Alexander
technique / written and illustrated by David Gorman. -- 2nd expanded ed.

Includes bibliographical references and index.
Also issued in electronic format.
ISBN 978-1-897452-75-2 (b&w).--ISBN 978-1-897452-76-9 (col.)

    1. Alexander technique. I. Title.

BF172.G67 2012   615.8'2   C2012-901163-0

First edition published in print in 1997
This second expanded edition (black & white version) ISBN 978-1-897452-75-2

Published 2012 by

LEARNINGMETHODS
PUBLICATIONS ™

Toronto Canada — www.learningmethods.com

LOOKING AT OURSELVES

EXPLORING AND EVOLVING IN THE ALEXANDER TECHNIQUE

Written and illustrated by

DAVID GORMAN

Also by DAVID GORMAN

– THE BODY MOVEABLE –
Blueprints of the Human Musculoskeletal System

– LOOKING AT OURSELVES –
Available in both black & white and colour print editions as well as in various ebook formats

This book is dedicated with love and appreciation to
Eillen Sellam
who not only translated parts of it into French,
but also translated me into France.

Now, some years later I have returned the favour
and translated her into a Canadian

LOOKING AT OURSELVES

CONTENTS

Introduction to this new edition

This new edition of the book contains all the material in the original 1997 print edition plus two new bonus articles never published in print before. Some of the illustrations have been improved and new ones have been added. In addition, one of the articles, *The Rounder We Go, The Stucker We Get*, has been updated (see details in the introduction to Chapter 7).

This second edition will be published in both black and white and in colour print versions, as well as in various e-book formats. The edition now includes an index and an updated appendix section with bibliography (further reading) and references with web links so you have access to the latest sources of information.

For the most part, the original material speaks for itself. However, it has been almost 15 years since the first publication of this book so some of the chapters need a bit of further explanation to place them in context. For these I have added new introductions at the start of the relevant chapters.

By the way, I am hard at work on another book about *The Anatomy of Wholeness*. This is a look at the larger whole-system patterns of our human structure and function from the point of view of the entire thinking, feeling, responding being. It will include a full exposition of my "pre-sprung elastic suspension system" model and its implications for daily life, for exercise and fitness, and for performance. The eight-part series called *In Our Own Image* in the book you are reading now gives a much shorter and earlier version of the territory of this new book.

Stay tuned for the publication of the new book, but in the meantime you can find further information about the content of the book and sign up to be notified about the publication date by going to www.learningmethods.com/revealing.htm.

In the meantime, I hope you enjoy this book, and find it useful.

David Gorman
Toronto, 2012

Introduction to the first edition (1997)

These articles and essays were written over a period of 12 years — a period of great discovery for me in my understanding of the Alexander Technique and how we work.

The first article, *Thinking About Thinking About Ourselves*, is the transcript of the F. M. Alexander Memorial Lecture I was invited to give to STAT (the Society of Teachers of the Alexander Technique) in 1984, just 4 years after I'd become an Alexander teacher. I had begun to have intimations of just how deeply our 'problems' and 'misuses' were bound up in our belief systems, our constructs — in short, our current 'reality'. Whole new vistas were opening up for me and I was very excited about the discoveries I was making. So I decided to talk about it all. The feedback I got was that very few people had any clue about what I was saying and, looking back, I'm not at all surprised.

Even though I can re-read what I said and see clearly in those insights the seeds of so much that I am doing now, the fact is it has taken me over a decade to realize and bring into practice the implications of the discoveries that were just forming then. These implications have exploded recently into fruition in a whole new understanding that has revolutionized my way of working and brought about a new work I am calling LearningMethods. This, however, is the subject for another day and I am 'retiring' to the south of France for a few years to develop it.

The next group of articles here, *In Our Own Image*, were planned as a series in the newly founded *Alexander Review*, which was published regularly during the four years from 1986 to 1989 and was the first independent journal for the Alexander Technique community. The series was my first attempt to commit to paper another discovery I had made about 4 years earlier. This one came out of my interest in understanding just what was happening in our systems when those so amazing Alexander experiences of sudden lightness and openness and ease occurred. I had already come to my first lessons in the Technique in Canada from a background of a lot of anatomy and physiology knowledge. By the time I had pursued this question through five years of lessons and three years of training I had a lot more knowledge and my own six hundred page illustrated text on the functioning of the musculoskeletal system*, but still there was something about those radical, changed experiences that I just couldn't fit into the usual notions of how our bodies worked.

One day in 1981 with my brain stuffed full of anatomy, physiology and kinesiology details, a new model of functioning came to me in a flash of understanding. I suddenly saw what was going on — a gestalt of how the whole system could in an instant be reorganized as an effortless spring-loaded unity. While this insight was there right away in its fullness, it again took me years (and many workshop of fumbling but gradually improving explanations) to express somewhat adequately in these articles. In the ten years since then, my understanding has deepened even further but that is for another book — it's not going to be much of a retirement, eh?

Initially these two discoveries — the power and depth of our belief systems and the

---

*   *The Body Moveable*, hardcover, originally published in 1980, is now nearing the end of its 5th edition (2002). The upcoming new edition may be issued in several versions: as 1 volume or in 3 volumes (possibly in a colour version too). More information is at www.bodymoveable.com

nature of our innate ease and wholeness — seemed to illuminate two different territories, but over the years they converged to form the foundations of further discoveries.

The next article, *On Fitness*, is an extract of a much larger entity which may or may not ever see the light of day. It originated as an on-going interview by Sean Carey. He had just published an interview with Walter Carrington and had become very interested in my work. We filled up dozens of tapes and had them all typed up into over 200 pages of manuscript, but hadn't got very far in editing it into a manageable size before his quest for what is the Alexander Technique lead him off in another direction.

I like the interview format. It's the closest thing we have to the reader reaching into the pages to ask someone the questions they would like to ask.

The following four short columns under the title, *Overview*, are not directly about the work, but are about our nature in an indirect way. They are about politics in the Alexander world — a boring subject for many. But what is politics really, if it is not about the process of making real all the conditions and structures that allow the things we hold dear to be born and flourish?

At the time I wrote these I was deeply involved in creating a new political structure that would bring together the various existing Alexander professional bodies and allow teachers who had previously been out in the cold to join the community. This meant bringing together the STAT and ACAT teachers (American Center for the Alexander Technique) in the forming of NASTAT (now AmSAT), changing STAT to allow non-STAT trained teachers to join and forging a mechanism to affiliate any new national professional groups. These column are part reporting on the progress of these events and part propaganda to help make sure that it would happen. In the end, after lots of negotiations and not a few fights, it did all come to pass, though it certainly hasn't turned out the way I'd hoped, but that's another story.

The second last article, *Experience and Experiments in the Alexander World*, explains itself in the reading. It is about an experimental master-class I gave at the 1991 International Alexander Congress in Engleberg, Switzerland. The Congress came at a time when I had been running my own training course for some years and my emphasis was shifting from teaching people about the 'Alexander' Technique to helping them get past techniques and back to seeing for themselves what is happening right in front of them — in fact the very thing that inevitably lead me away into a new work.

The final article, *The Rounder We Go, The Stucker We Get*, comes from another period than the others items here. It is closer to where I am now than where I was then. I won't say much about it except that as part of the journey, it helps show today's direction.

I hope you enjoy this book and find some value in it. I am always happy to hear from any of you about any of your responses to what I've written. By the time this book will be published I'll have moved from London without a new address, so the best way to reach me these days is by e-mail (david@learningmethods.com) or via my website (www. learningmethods.com/contact.htm), otherwise you'll recognize me in the south of France as the one with the Canadian accent and a computer under his arm.

David Gorman,
Provence 1997

Chapter One

Thinking About Thinking
About Ourselves

THE F. M. ALEXANDER MEMORIAL LECTURE
delivered before
The Society of Teachers of the Alexander Technique (STAT)
on October 27th, 1984

Welcome! I want firstly to thank the Council for the honour of being here to give this talk, and secondly to thank you all for coming. What I want you to join me in playing with tonight is how our ideas affect our physical use; how our thinking affects our bodies. I want to explore how we think — particularly the aspects of thinking that involve our beliefs; how our individual beliefs tend to organize themselves into a system of beliefs and how these belief systems then coalesce to become a 'reality', producing for us a relatively self-consistent overall perception of ourselves, the world and our manner of living in it.

We'll then look further at how different belief systems constitute different 'realities', and how these different realities lead into their own correspondingly different worlds of experience. Our belief systems, that is, tend to imply certain patterns of use. It is inescapable once we operate from the basic premises of certain beliefs that we will tend to have a way of use in accordance with them — be it a poorer use or a better use. In other words, we use ourselves the way that we think of ourselves. Of these realities, in still other words, some are more constructive than others.

Let us begin with some very general cultural beliefs… We are creatures who believe that we live on a rather large planet that is whizzing around in a rather large universe. We believe that on this planet there are a variety of 'things'. Some of these things appeared on the planet without us having very much to do with them. Some of these things we actually put together and made out of parts of some of the other things that were there without us having very much to do with them. We believe that we move about on this planet among all these things making a living, doing activities, and responding to events. We believe that we 'have' bodies which 'belong' to us and which are separate from other things since they come along with us when we move among the things in the world. We believe also that our bodies have different parts with different functions; some of which we can control and some which we can't.

It is not difficult to find, in our culture, subscribers to these beliefs. We believe in the system not only because it works, and is confirmed, for us in our lives, but also because a lot of other people tend to believe too. An immense amount of power is carried by the system because it's so common; a sort of 'common-sense' pervades and renders it true. But is it so true? And, maybe more importantly, does it really work so well?

We believe that there are stars and a moon out there. We believe that the sun comes up every morning… Ah, but here we've got something interesting! We all know the sun comes up every morning; we experience the sun coming up every morning (except, of course, here in London where the clouds get there first). But, on the other hand, we also 'know' that the sun doesn't really come up; rather the earth turns and each day we are brought around to a new morning. It is obvious that these are different ways of looking at the same thing; different 'points of view' of the same phenomenon.

Another example of these different points of view is how we find it very easy to speak of 'taking a breath', thinking of it as sucking in air. Yet at the same time we know that what we are 'really' doing is opening a space inside ourselves for the pressure of the atmosphere to push the air into us. Our deepest beliefs tend to be based on our own experiences. We'll see as we go on how very important it is for us to become aware of the implicit point of view contained within any particular belief system.

The difference between the point of view, for instance, of the sun coming up and that of the earth turning (or of sucking in air versus it being pushed in) is the difference, respectively, between 'subjective' and 'objective'; the difference between us 'experiencing' something and 'knowing' something. We experience the sun coming up in a direct sensory way while we don't actually experience the earth turning. The fact that the earth does turn is an objective point of view. It is a point of view from outside ourselves; a bird's eye view; or better yet, the universe's point of view. The experience of the sun coming up is more subjective. It is a point of view as we see it from within; from the information of our own senses; from our 'still point at the centre of the universe'. Both are, of course, different, but equally valid, ways of looking at it. Each has different implications and different uses. Of necessity, each takes us into different modes of thinking and experiencing.

With this in mind I want to dig into some of our fundamental beliefs about our existence on this planet, and look at the 'givens' behind them. By 'givens' I mean that which is given; the phenomena which are just there, in and around us all the time (of which: 'sorry, nothing you can do about it; just happens to be the way that it is'). We'll see how our point of view shapes and shades these givens, channelling us to respond within the framework of the belief. This structuring occurs such that fundamental and basic beliefs about the nature of the givens provide the foundation (and the architectural style) upon which further beliefs about our own nature are elaborated, which in turn imply still further beliefs about using ourselves.

We live on the surface of the Earth. The world is rather large compared to us, and one of the most obvious, taken-for-granted aspects of living on Earth is a very commonly held belief in this thing called gravity. Gravity is a name that surrounds and holds within it an experience that we all have. It's a point of view and a framework for describing the relationship between ourselves and the planet. Using our beliefs about this relation to the ground as a foundation — a very appropriate place for a metaphorical foundation — we can trace the implications of these beliefs to see what sort of superstructure constellates around them, and what are the results of living within such a system.

When asked, most of us will respond that the biggest attribute of gravity is that it 'goes down'. Thanks to Newton, we all know that it is gravity's fault that things fall back down when you throw them up. We tend to think of it as a sort of omnipresent force all around us which is constantly drawing things downward to the ground, rather like a steady and never-ending drizzle. The implication of this downward force is that gravity gives us 'weight'.

Now, we as creatures are built of many different parts quite closely, but loosely, connected with each other. That is, we are flexible and moveable creatures. We as humans, in particular (compared to the other animals), are very unstable creatures. We have so much of ourselves so high up off the ground over such long moveable bones that our 'weight' is constantly threatening to fall because of the down of gravity. Within the terms of this belief system we will find it very difficult not to have gravity, weight and instability conspire to make us do 'effort' in order to deal with them.

Does it make sense to most of you that to do this effort, which is needed to deal with our unstable body-weight in gravity, we have to use our 'muscles'? Do we not think of our muscles as the 'active' parts of us? They are the parts we use to get hold of our rather inert bones to keep ourselves from falling. In other words, to hold ourselves up. If we let go of ourselves we fall down. Of course, what I'm describing here is the majority belief-system in this culture. It will not necessarily be what we each believe individually. It also seems self-evident to most of us that most 'movement' doesn't take place without muscles doing effort — effort that they do by 'contracting'; by working as if they were a whole group of active little hands that grab hold of our bones to keep us upright or to pull us into activity.

Well, so far so good. Most people (you know, the mythical *Mostpeople*) will think that the above makes perfect sense — "I mean, everybody knows that!" In terms of our metaphor, we've come up from the foundations in the basement to the ground floor where most of our daily life takes place. Company is plentiful here, however, if we follow the implications just a bit farther, we can see that this set of beliefs has already created somewhat of a problem for us.

When we hold our unstable selves up in gravity by using contracting muscles, we can't avoid that fact that we are 'holding' ourselves. This is not just semantics here, but a very real physiological event. Consequently, we've got ourselves in a bit of a bind (pun intended). We need to be able to be upright to move around in the world and do our jobs; yet we are getting that uprightness by holding ourselves. In addition, we all have a desire to be as free as we can be. Yet, through gravity, weight, effort, muscles, and holding we've arrived in a situation where we get our 'upness' by holding onto ourselves. We're either up and holding (that is, postured and not very free) or, when we let go the holding to free ourselves, we collapse and go down. We can either get the up at the expense of the freedom, or we get the freedom at the expense of the up.

There are not many ways out of that bind without a great deal of confusion entering into the attempt to explain how one can go about freeing and still be upright while doing so within the terms of gravity, weight, effort, and so on. We have, in effect, a contradiction in terms. As long as we stay operating within these terms, this conflict becomes difficult to resolve without becoming overly simplistic, vague, or somewhat mystical.

So what is the significance of all this? With this way of organizing ourselves we are operating in a situation where we are constantly in a struggle — we are *fighting* gravity. Gravity is the bad guy who gradually and inevitably pulls our tissues downwards, and eventually drags all of us six feet under the ground. When we think of gravity as a force all around us with the attribute of having 'down-energy', we are making the relationship between ourselves and the planet into an abstract concept; we are objectifying it. By conceiving of it this way, from a point of view outside ourselves, we constantly put ourselves in the position of having to do something about it. Our beliefs are structured so that we have to react to gravity to achieve what we want — hence the conflict.

The significance here is that we are not really fighting the abstract of 'gravity' at all; we are fighting ourselves. If we are holding ourselves up (getting a grip on ourselves) by means of muscles contracting, then we are organizing ourselves with a way of operating based on contraction — we are contracting ourselves. We stay upright and move around by getting hold of parts; pulling and levering them around other parts. Our muscles then work by shortening, by squeezing our bones closer together and pressuring our organs. We end up hanging onto our skeleton for dear life! And there's not much freedom in that. Those of you who are Alexander teachers, and probably even those of you who have had lessons in the Technique, can recognize this essential approach in the problems with which people come to this work — those common habit patterns of either getting hold of themselves and tightening into themselves or of slouching and collapsing when they 'relax'.

What we have done with this point of view is to objectify ourselves. We've made ourselves into an object — a series of falling weights that we have to hold up. We have to use one part of ourselves to do something to another part of ourselves. This is the essence of the mind/body split. We have an objective part of ourselves, our body, acted upon by gravity while the subjective part, our consciousness (the little man behind the TV screens up in your brain), is monitoring and operating your body. As a mode of approach, this way continually shrinks our 'self' inwards, from our bodies, up into some little point of consciousness inside our brains. In our metaphor, we have now moved up the stairs to the first floor above the ground level where, in the privacy of our bedrooms and bathrooms at the end of the day, we admit of our tensions, our symptoms, and wonder what's going wrong.

If our 'common-sense' beliefs lead us directly toward our problems, what do we have to do to resolve this seeming paradox? The Alexander Technique teaches us effectively in practice to break out of our habits to a more free, more open, and 'up' way of use. We might ask here, what is so different in the way the Alexander Technique gets us using ourselves? How is it such a different way of thinking and operating than most people are used to?

Let's go back to the basement, to our relationship with the earth, and see if a different point of view will get us a different 'reality'... The planet is poised in space and we dwell on its surface. From a more subjective point of view what gravity is about is that no matter where you go the planet will always follow you around coming right up underneath and supporting you. No matter what you do (in normal activities) you can count on being supported.

Let us use a different name to encompass this different aspect of our relation to the planet. We can change Gravity to Support. There is a far bigger difference than just a name here. Gravity for most people is an enemy; support is your friend. We need support and if we can learn to use it skilfully then we have a very powerful tool at our command.

When we allow the planet to support us (or allow ourselves to simply rest on the planet) as we move about, we get quite different results. Instead of experiencing the struggle and effort of reacting to the abstract of the 'force of gravity', we directly experience something much more tangible — the planet itself. Our point of view now spreads out from ourselves to include the planet and our relation to it, instead of shrinking us back into a small point of consciousness. We now perceive 'gravity' as our in-the-moment contact with the ground, and as a bonus we instantly perceive our changing relation to our support as shifts in this contact. From the other point of view we had a system of concepts in which we constantly had to react in order to avoid something we didn't want. Now we have one where the more we use it, the more we get what we need. The weight and effort resulting from the other point of view are also very tangible experiences; but they stem from an approach that is not nearly as constructive in practice. So, down in the basement we've found there are two doors; each leads to a very different space. It's up to us to choose which we enter.

We will carry on deeper into this in a moment, but first to give a little concreteness to all of this I'd like to ask your indulgence to play a game. There's an exceptionally full hall tonight, but I think we've got enough room to do this. I'd like you all to stand up, find someone nearby to work with and face them. Now I want each of you to come to a way of standing where you have a sense on the bottoms of your feet of a more-or-less even distribution of contact with the floor — as much contact to the front of your feet as to the back; as much to one side as to the other. Then one of you bring up your two hands in front of you with your palms facing each other and all your fingers pointing up to the ceiling. The other person bring up one of your hands and place it right in between your partner's two hands so that your fingers are pointing up too.

Now, the person with the two hands up, quite simply bring your hands together onto the other person's one hand until you sense on your palms and fingers roughly the same amount of contact as you feel on the bottoms of your feet. I want you to do this fairly quickly; take just enough time to go: 'Hmm, yeah, that's sort of about it; no, that's too much; now that's too little; yeah, it's sort of somewhere around in there'. When you've got your hands and feet feeling about the same pressure, let the other person know and both of you just take a little snapshot in your memory of what it felt like, then switch and do it the other way around. When you've all had a turn, I want to make a statement to you and then ask you a rhetorical question...

(Several minutes pause here for people to do the experiment...)

The statement is: 'What I asked you to do was to recreate on your hands your subjective sense of *your entire weight*'. The question is: '*Did it feel like it?*' You objectively know that you weigh a hundred-and-some pounds. You can stand on the scale and read it off. Did it feel like you were squeezing with or being squeezed by a hundred-and-some pounds?

(Everybody answers: "*No no no no no*")

Did it feel like twenty pounds?

("*No no no no no*")

Like five pounds?

("*No no yes no yes*")

Three pounds?

("*Yes yes yes yes yes*")

We have here a rather large mismatch — A mismatch between what we objectively know and our kinesthetic reality of the moment. What your senses are actually telling you is that you are just very lightly resting on the planet. What happened to all that 'weight', and why are we doing all that holding and gripping when we are just lightly resting there? As you can see there's quite a broad gulf between the implications of these two points of view.

It's very easy for us to 'know' that we weigh a certain number of pounds because we have these little measuring instruments called scales to tell us. I stand on it and it says, "one hundred and forty". If I was to walk over and pick up a sack of flour labelled "one hundred pounds" (which is lighter than most of us), then compare that hundred pounds with my even greater 'weight', I will naturally think: "Gee, am I ever heavy." It's not difficult, once we make that correspondence, to begin to think heavy and start to feel heavy. We begin to move heavily and go around carrying, literally, hundreds of pounds through the world in our daily life. (No wonder we want a good collapse now and again.)

But there is a very big difference between a sack of flour and you. One is something that is outside yourself which you are using yourself to feel; the other *is* yourself. *You* need to hold up the sack of flour, but *the planet* will hold you up. We have, in effect, sensitive scales in our feet to reassure us of its support. The reality that your feet are telling you is, in a sense (another pun), a much more direct reality than the abstraction of pounds on a scale. We can learn to deal directly with this tangible support, allowing ourselves to bring a sense of security and lightness into practice, but only if we choose to direct our attention to it, not becoming seduced by how much weight we 'really' have and how much effort must be necessary to keep us up.

You can all sit down now… When you were standing there and as you sat down, were you aware of being lightly supported as you began to move? As you were moving? Even as you were touching the chair? And as you let the chair lightly support you? Or, did you drop back into the chair, feeling your weight and the increased effort to control it, only now becoming aware of support and contact? This necessary security of supportedness is an undeniable reality that is, at any moment, accessible to us. All we have to do is look for it. Then, and only then, do we have a chance of using it as a basis for movement. Remember also, that what you're feeling on the bottoms of your feet is *all* of you on the ground. No other part of you has all of you over it, so every other part of you could be used even more lightly than what you feel from your feet.

Let's go on to some further implications of that belief system which has lead us to weight, effort, and holding. What is the objective of this holding up? That is, what constitutes successful 'up'? Is it just that we don't fall down? Obviously not, since we also want freedom served with our up.

Well, it doesn't take too much to figure out that the less instability we have, the less we will tend to fall, the less effort and holding up we will need, and the more freedom we will have. So, naturally, we will be looking to get our collection of unstable weights as much up over themselves as possible. In other words, we'll be concerned with 'balance' — the sort of balance where perhaps, if we could stack ourselves up over ourselves (like piling up building blocks), getting it just right, we could let go the holding and become free. Thus, as soon as we become concerned with balance we also become concerned with 'alignment', and when we're concerned with alignment it's difficult for us not to start to pay attention to 'positions'. Am I straight? Am I vertical? Have I got myself sort of up over myself? Is this *good* alignment?

This sounds eminently reasonable if you imagine those stacked-up building blocks again. However, we are not a set of blocks with our parts symmetrically arranged or symmetrically moveable on each other. Because of two structural 'givens', it is not possible to balance our human bodies in the same sense that you can stack up blocks. The first is the simple fact that we are alive — we move and breathe, our hearts beat constantly, upsetting any static balance. Thus, the best we could hope for would be a dynamic, constantly-reaffirmed balance with our muscles forced to grab us when we go off balance and then pull us back to alignment. While it is possible to 'stack up' your leg bones in this kind of dynamic balancing, even though they keep wanting to go off balance, we generally find it easier to lock or hold our legs (quite often in an alignment which is not even remotely balanced).

Your torso, however, is a different story. There are entirely different kind of joints in your spine — what I call distortion joints. Your discs are flexible elastic structures, meaning that there is no movement in your spine without those discs being distorted in some way — either squashed, bent, twisted or stretched (with no particular connotations at this point about whether these distortions ultimately are 'good' or 'bad' for your discs). Your torso is also inherently unstable in a forwards direction, as we all know. When we get tired of holding ourselves up, we start to slouch out forward. To put it in different words, there is more of you in front of your spine than behind your spine. We cannot 'balance' our torsos in the sense of getting all its parts up over each other without pulling ourselves (using grabbing muscles) up out of that inherent instability and then holding ourselves there against the elasticity of our now distorted discs.

In the face of this elastic and unstable liveliness of our bodies, whatever 'alignment' we do achieve is going to require constant and fine adjustment. Naturally, we will want to get our balance as well-aligned as we can, then, hopefully, keep that good alignment where we've got it while we see if we can get a little bit more. To the degree we are successful at this aligning, we will end up deviating from this 'right posture' less and less, hence allowing less and less flexibility, until we get into a position where we hardly move at all anymore. We are no longer poised, we're postured.

As a matter of fact, as you're all aware in yourselves and the people you work with, it doesn't take long before the range of deviation from the good alignment in which we carry ourselves becomes so small that it is smaller than the range of flexibility needed for free breathing. In other words, we hold onto our breathing so it doesn't disturb our 'free' balance. Huh? This seems like pretty strange territory to end up in, considering that our path began with beliefs that made so much sense in the beginning. The more we follow this path the further away we get from what we want; and the more confusing it becomes to try to get what we want and discover that we keep getting something else!

Time to go back down to our structural 'givens' — one pertains to the game we played earlier. Built into you is a very powerful tool for recognizing and coming to support. As an upright creature, a very unstable upright creature, you so happen to be built that when you rest on the planet in such a way that your sense of contact is more or less evenly distributed on the bottoms of your feet, you are directly over the planet and it is supporting you totally. This means that within that range of contact you cannot fall. That is, all of you cannot fall since you are already on the ground and there is nowhere lower than the ground to fall to! This simple evenly-distributed contact is directly tangible and very easy to find. All you have to do is look for it and go there. You then know that you've taken care of your major security in terms of support on the planet and a base for movement.

All of you can't fall when you are directly over the ground. However, it is conceivable that part of you could fall off another part of you. This toppling over is what we usually mean when we speak of falling down. But there's another 'given' that takes care of that one. We are so built that it is not possible for a part of you to fall off another part of you unless you give permission for it to happen. Not only do you have to give permission for it to happen, but you have to give very active permission. And, better still, the more that you begin to topple, the more active permission you have to give to allow it to continue. Most of us, especially on a floor like this, won't give that permission more than just a little bit.

Putting this all together, perhaps we can get a little closer to understanding the strange territory we were in a minute ago, where the very way we try to be free gets us more tangled up in holding. If it's so easy to know when we're over the ground, and if (short of tripping or stumbling) we have to give permission to fall, then what exactly is going on with this holding up stuff anyway?

It is inevitable that once we force ourselves into habitual holding we will begin to feel that holding. Most people don't know exactly *how* they're holding, but after a couple of hours they can feel the soreness, the tension, the stopped breathing, whatever are the symptoms of that holding. It's here that we can see how important our point of view is. For, if you believe that you are holding yourself up and you let go that holding to free yourself, where is the only place you can go?

Down... It's built into your way of approach. You let go holding up, you'll come down. Every time you 'relax' your holding, thereby losing your uprightness, you will have your belief system reaffirmed. You'll say: "Ha! See I told you, I have to hold myself up because if I don't I will fall down!" Right there is the permission we have to give to topple over. It's implicit in the belief that if I don't hold myself up I *will* fall down. So when we stop holding ourselves up, we drop. In fact, what we're really doing when we make parts of us into weights that other parts of us are forced to hold up, is dropping ourselves and then holding up the dropping. We must be still dropping while we're holding up, because if we weren't, we would have nothing to have to hold up — another conflict in which our way of thinking can tangle us.

Hang on, it becomes curiouser and curiouser... If we return to our structure again with a different point of view, we find that even when we think we're holding ourselves up, we're not really doing that at all. What we're doing is holding ourselves down.

The fact is that there is more of you in front of your spine than behind it; in other words, we are unstable forwards. When we turn off our holding up, the upper part of our torso slouches down forwards with our lower back and hips slouching backwards. When we haul ourselves back up again, we do so looking for the result we want — to get up. We pay attention only to the end we are trying to gain and don't really notice what we're actually *doing* to get there. There are no sky-hooks up there, of course, to grab onto and lift ourselves up, so the only way to pull ourselves up is to use that powerful set of muscles that runs up and down the back. These muscles have to pull down on our back in order to lift up our front and squeeze our lower back forward. They then must keep on holding us down to keep on holding us up! So the actual muscle work of it, the actual doing, is a pull *down* the length of our back. Thus, we see that 'weight', in any experiential sense, is a self-created phenomenon — we weight ourselves with our own muscles by pulling down on ourselves.

Until we can reveal to ourselves the 'doing' side of it — the actual pull down — it's inevitable that we are going to be stuck in our point of view, and consequently stuck with our habit of holding up and all the symptoms that go with it. Now, if we can start to catch ourselves in our pulling down, and we can release that pulling down, where can we go?

Up... Quite a different direction — approximately 180 degrees different — from what we normally think of as gravity. Hard to imagine how one could let go upwards —

it seems to defy that whole other belief system. And yet we got here by picking at a few loose thread on the 'reality' and finding that we've unravelled a line of reasoning which, at the very least, gives us a different perspective to work with. As we begin to catch ourselves in these 'mismatches' between systems, we begin to have the opportunity of choosing a point of view which might actually lead where we want to go.

If we can catch ourselves at the actual doing of holding ourselves down and manage to release that holding, we go up. We are then, in the deepest way, changing our whole way of organizing ourselves so that, instead of trying to *hold* ourselves up, we can get up by *releasing* our holdings-down. In other words, we can use our muscles to let go of parts into activity — into more freedom, more openness, more length, more breathing, and more flexibility.

When we ungrip to let ourselves go up and open out, we allow an expansion to come in. That's the opposite of the contraction and inward pressure we saw before. The expansion allows more freedom of movement and more sensitivity, not only inside ourselves, but also outward to the planet and our daily life. This increased sensitivity makes it easier to notice our changing supportedness and, also, to respond accordingly. When we are weighting ourselves and holding up the weight, we tend not to be aware of where our support is — we aren't looking for that kind of information. Instead, we're looking for alignment, hence, we can't use our sense of lightly being supported and are stuck in the vicious circle of balancing weights.

It is very difficult, when you are subscribing to one belief system, to allow the elements of another system to come in — they tend to exclude each other. A belief system will always tend to expand outwards until it equals 'reality'. Operating within a system gives you corresponding sensory feedback such that you experience yourself and the world in terms of that system, which in turn corroborates the whole approach of the system.

Let me give you another example of this. There is a lot of attention these days in our fitness-conscious, beauty-conscious, thin-conscious, culture focused on our abdominal area. We generally have a conception that all our organs are just going to come plopping out if our abdominal muscles don't do their job of holding everything in. So we work hard to get those flabby muscles trim, toned, and strong so they can keep everything nicely in there. The usual way to achieve this is through 'strengthening exercises' such as sit-ups, leg-raises, rowing exercises, and so on.

Structurally speaking, these muscles stretch between your ribs and your pelvis. The major set of muscles run in a crisscross fashion diagonally, while others run up and down from the front of your chest to your pubic bones. When we work hard at these exercises, we are practicing getting very good at shortening our abdominal muscles; we're getting very strong at pulling our ribs closer to our pelvis; at pulling our ribs *down*. We are training ourselves to use these muscles to reach down from our ribs, grab hold of our organs, pull them up and in, and then hold them there. We are hanging our organs from our ribs — in essence, hanging weights from our ribs. Since the crisscross muscles narrow our chest when they shorten, these pulls act not only downwards, but inwards as well, pressuring our organs. No wonder they want to pop out the moment we let go — it's not in spite of the 'strength' of the muscles, but because of it!

Strangely enough, moreover, it seems to be that our ribs have a lot to do with breathing. Breathing seems to be the sort of thing that has to do with expansion — an expansion where our ribs open upwards *and* outwards all the way around us. Thus, if we do any abdominal holding-in, we're using our own muscles to interfere with our breathing. In addition, it is inevitable that if we do any contractile 'effort' with these muscles in front, we also force ourselves to do similar effort in our back. Our spines are, after all, flexible structures and if we pull down with these front muscles without any compensatory pulling-down in back, we simply bend ourselves over forwards. So now we have two sets of muscles pulling down on us — and we wonder where our heaviness and tension comes from?

Here we are with a skeletal structure, with its connecting ligaments and capsules, that is very free. There's absolutely nothing in your skeleton to stop free movement. The only thing that can stop us from moving freely in ourselves is that our muscles *won't let go* of our bones. As soon as we start any grabbing on with our muscles to shape or posture ourselves, we will have to begin compensatory holdings elsewhere until the contraction spreads all over us. (Each person's patterns of holdings and droppings, tightenings and squeezings, will naturally vary according to their own ingenuity and determination.)

In a manner similar to holding-up posturally, if we notice our *abdominal* holding (up and in) and then let go of the holding, of course our organs are going to drop out. This reaffirms the need to hold them in, and around we go again. But if we sense how we are holding ourselves down and, instead, let go of our *ribs* upwards and outwards, we simply allow ourselves to breathe again. We give ourselves more space; our organs are happier; our muscles lengthen and are more 'elastic'; and our in-built breathing reflexes are freed and activated — all because we stopped interfering.

Let's come back to this concept of balance again. There is another problem we create for ourselves. When we search for good alignment, we tend to gradually freeze our liveliness and flexibility into 'right positions'. As we become stiffer and more stuck we begin to have a sizeable amount of inertia to overcome when we go into movement. That is to say, it becomes easiest for us to go into movement by actually going off balance so as to get the momentum of our 'weight' working for us. Walking, for instance, is described in many texts as a "continually-arrested falling" — we lean forward off our supporting surface area then force ourselves to react by catching ourselves.

This is very obvious in sitting and standing also. It is astounding how so many people sit down by heading blithely back off their feet to begin. As they leave their supporting surface little stiffenings increase in their necks and backs; tension starts in the front of their thighs; their arms may reach out; their toes pick up a little off the ground and you can see the tendons in the front of their ankles stand out. All these events aren't extraneous habits we have picked up, they are balance reactions — the sort of things which happen when you go backwards off balance.

Similarly, if I pose the question to a group of people: "What do you have to *do* to stand up out of a chair?", they will tend to describe a combination of the following: "I have to lean forward", or a little more actively: "I have to thrust myself [or pull myself] forward". "I have to then push down [or push myself up] with my legs to get up". (Everybody knows you have to push; how can you get that *weight* up unless you do some *work*?) "I have to lift [my bottom] off the chair in some way". (They will probably not be aware of all the tightening in their neck and the arching of their back during this sudden 'oomph' of pulling off the chair.) "I reach out with my arms as I get up", or sometimes: "I put my hands on my knees and push myself up". (This is an interesting one — pushing down on a part of yourself in order to lift another part up!)

From a different point of view, these things we experience as essential parts of standing which 'I have to do in order to get up' aren't things we are 'doing'; they are simply things we have become accustomed to feeling because every time we try to get out of the chair before we are remotely over our feet. All of the above efforts — the tightening of our neck, the arch of our back, the reach of our arms, the grab in our thighs — are not really actions. They are the balancing reflexes of us suddenly having to grab ourselves from falling and reach for balance because we aren't over our new support (our feet) before we try to leave our old one (the chair). We just get so used to them happening that we think we're doing them. Interestingly, there's nothing like going off balance, then grabbing and lifting the off-balanceness, to give an experience of weight and effort. Thus we become convinced that we have to do quite a bit of helping out to get our heavy old selves up off that chair.

Of course, in a sense, we *are* doing all those efforts. We *force ourselves* to do them because we have almost no awareness of where up-over-the-ground really is. We don't have any real experience of balance and support — all we've got is a series of experiences of reaction when we've gone way off it. We have no real set of experiences of what it's like to use gravity intelligently and skilfully. And one thing the Alexander Technique is about, is learning a skill. At least on the physical side, we refine a skill of using ourselves in the world, in whatever activities we are doing, in such a way as to free and open ourselves (not only the self we are at that moment, but to all our as-yet undiscovered possibilities).

There are so many different ways we can move; different shapes we can get into; and different activities in which we can partake. We are such free structures with so much potential. If we don't have the skill to manage all that freedom, we still have to go on living and functioning, so our only recourse is to shut down parts of that freedom in order to be able to manage it. We generally accomplish this by using our muscles to freeze joints by gripping our bones until, for practical purposes, there is no joint there to manage. We end up reducing the possibilities until there become only a few places of movement; few enough for us to deal with. Alexander's discovery of our constructive central organization, and his technique for getting across how to access it, teach us how to gradually discover for ourselves the skill of letting ourselves be freer, lighter and more open — and the further skill of staying with (or keeping coming to) that new central organization so that, while being freer, lighter and more open, we can learn how to go about our various activities.

Well, so far tonight we've explored how our thinking affects our use, and the implications of applying a more constructive point of view to the fundamental 'givens'. Let me begin to finish by using this tool to climb up into the attic and look at our thinking itself.

Our brain (that is, our forebrain) is built in two functionally somewhat different halves. The left brain is primarily concerned with analysing. To use our point of view of subjective/objective, it is the objective part of our brain. Processes in that half attempt to deal with movement by breaking it down into components and directing parts individually. That mode of thinking is very well-suited for directing activities out in the world. That's the part which can figure out how things work. That's the part which conceptualizes, gets insights into mechanisms, sees patterns in things and connections. That analytical mode works beautifully when focusing our attention outside of ourselves, using these insights and ideas as a guide, to recruit and direct our parts in the manipulation of objects, in making things, and in invention. That part of us is not very well-suited to dealing with directing us as a whole in postural 'activity' or in movement because it inherently wants to break it down into bits and direct us as a series of parts. That mode is very linear, very cause and effect, very 'objective'. When we operate that way on ourselves, we inevitably start to lose our integration and sense of coordination — we end up splitting ourselves and losing the whole.

On the other hand, the right side of our brain 'experiences' movement rather than analysing it. Through that area we experience as a whole. This is the part of your brain which is sensitive to your kinaesthetic feedback. This is the part of your brain which can perceive support; which can perceive openness, and freedom, and liveliness, and all the incredible amount of information which we need in order to use ourselves skilfully. We have no hope of being able to use ourselves with a highly responsive skill without a great deal of reliable information. If we don't have enough information available to us, we have no choice but to shut down possibilities and respond with a relatively crude level of activity in whatever we are doing. Thus, when we embark on a process of integrating and unifying ourselves; of gaining a sensitivity and an appropriate responsiveness; of freeing and opening ourselves we are largely developing our right brain functioning.

However, it is the left part of our brain which handles verbal explanations; which figures things out; which is able to talk about things. We are going to have a bit of a problem if we successfully allow ourselves to experience ourselves and the world around us as whole, and then try to analyse it and talk about it, since our left brain didn't *have the experience* and so can't really explain it. If we attempt to explain it in left brain mode, we'll have to reduce the experience down to something that is figure-out-able. This abstracting, conceptualizing mode of consciousness is only part of us — there is a lot more happening in us than this part can isolate and focus detailed attention on. This 'objective' mode is like Procrustes when it encounters a new experience outside its territory. It will encompass and explain the whole by chopping it down to fit its present concepts and terms. When we do this, we, as a whole, shrink a little bit along with our thinking. As our point of view narrows down to concentrating on fiddling with the parts, our muscles also contract us, pull us down, narrow us; and at the same time our experience of our *self* shrinks up into a point somewhere behind our eyes. When we use the products of that reductive thinking in communicating with other people, in stimulating their thinking, and encouraging their potential, our effectiveness will diminish proportionally.

This doesn't mean, though, that we can't use words constructively at all with this process, because obviously I'm doing it right now. We need to learn to use the language, not as an end in itself to model or 'explain' the experience, so much as to prepare us and lead us up to the experience. I think the best use of language is to lead us into a point of view where we can catch the mismatches of our belief systems in action; where we can reveal and understand the premises of our habits. We need also to explore and/or expand the repertoire of language for more constructive concepts and phrases which more closely mirror the quality of the experience we discover, so that in thinking and speaking it, we are greasing the way to living it.

This means allowing our thinking to lead us to inherently paradoxical places; somewhat circular places; places where we don't need to immediately try to figure out and resolve everything. After all, hopefully, they are new and unknown experiences — a little larger than our 'old' selves.

Let me emphasize here that I'm not disparaging the objective point of view. It is, of course, a point of view that describes the way things work when considered objectively, i.e. from outside ourselves. However valid (and valuable) it is when directed outwardly (and that includes looking at the human body as an object), it ceases to be constructive when turned inward and used on our own selves as a basis for activity. That left brain mode of understanding only leads us right back into the sort of use and experience which we are trying to get out of. For, we cannot figure out this new reality — the only understanding is the actual experience itself. After we've allowed ourselves to re-organize into a new experience of activation, of opening ourselves, of release and lightness, it all begins to make sense, and we can say: 'Ah, now I see what you mean!'

Our mind and our thinking can be a very powerful tool if we can gain command of our attention and learn to direct it constructively. When we learn to use that tool appropriately, our language and the way we use it can engage people's thinking in such a way as to facilitate them letting themselves have the experience. As a result, they will find, gradually, that the point of view of this new pathway (the means) will be reaffirmed by the understanding that comes with the actual experience itself.

In other words we have to become comfortable with that left brain part of us feeling a little hungry, a little unsatisfied, while we stay close to the experience and avoid abstracting it. That left brain part likes to consider the right brain contribution as only supplying 'raw material' of sensory experience which it then has to polish into a finished product, nicely wrapped in meaning. We need to respect the reality of the experience — its awesome depth, its emotional scariness, its open-ended newness. We have to embrace the seeming paradox of getting what we want by giving up what we have. We have to acknowledge those mismatches — those parts which don't fit — between the old and the new and just let ourselves exist, for the moment, as two different 'realities' in the same place at once. There is an immense amount of energy-of-discovery generated by the conflict of two different realities in the same place at once. The inevitable resolution of that conflict will occur by itself. Our job is to create and maintain the forum where that working out can take place for us. It is an extremely constructive place where we will grow and change and become more free. In other words, it's a process of evolution which we can choose to enter at any time.

That's really all I wanted to say…

Thank you.

Chapter Two

In Our Own Image

Introduction to this Series

Publication of this new edition marks 30 years since I first made the discoveries that led to this series of articles. That was in 1981 and was the culmination of a quest I had been engaged in since the late 1970's. I had been trying to work out just what was going on in people's systems, anatomically and physiologically, when they experienced that dramatic and sometimes sudden physical change from their usual tension, holding and effort to a state of freedom, wholeness and lightness.

Try as I might, all the explanations from my existing anatomical knowledge didn't really seem to cover the depth and profundity of the change. I realized that I had to go back and rethink the whole way we are built and how we function. So, during the next three years (early 1978 to late 1980) I researched all the state-of-the-art knowledge about muscle structure and function, joints, posture, kinesiology, and so on. My "notes" for this exploration eventually turned into my first book, the 600 page, illustrated text, *The Body Moveable*, now in its 5th edition.

Then one day in 1981 — even today I remember everything about that moment exactly — it all came together. I had been reading T. D. M. Roberts' detailed and comprehensive book, *Neurophysiology of Postural Mechanisms*, and a phrase on a page[*] just jumped out and everything fell into place. I had just grasped, in a flash of insight, an entirely new understanding of how our system can work as an integrated whole — the *pre-sprung elastic suspension system* I've since come to call it.

I couldn't explain it very well at first, especially since the gestalt of this model is hard enough to visualize let alone articulate, though dead obvious when experienced. Nonetheless, the vision was clear and bright in my head, and years of teaching gradually gave me better and better ways of getting it across to others.

In the beginning I was bowled over by this discovery but still not sure whether it was truly new on the one hand and on the other hand whether maybe I'd misunderstood something important. As time went on, however, it was clear that it was new and revolutionary to everyone I taught and of all the medical and engineering people I ran into no one could pick any holes in the science behind my theory, particularly after they had experienced it too and appreciated its implications.

What was especially nice for me, after discovering and developing this on my own, was to find that some others had been thinking along convergent lines. For instance, some years after my initial discovery, one of my students sent me an article by a Rolfer, Ron Kirby, called "*The Probable Reality Behind Structural Integration*". It was about the tensional arrangement of soft tissues in the spine. A lot of Kirby's ideas were similar to mine though he

---

[*]   It was in this sentence on page 164 (2nd edition, Butterworths 1978): "If no other leg is available to support the weight during the reflex step initiated in the way just described, the *upthrust is suddenly increased to throw the weight upwards* momentarily…" [my italics]

It was the word "upthrust" that made the light bulb go off and in that moment I had a *gestalt* of the whole mechanism. Nevertheless, it took me some years more to be able to adequately explain it to others and I am still recognizing new aspects of its implications.

15

restricted his "gravity supports the body" idea largely to the structural tissues around the vertebral core whereas my insight is a more global model engaging our whole system at all levels including the level of you the choosing person who can interfere with the working of this suspension system.

What really grabbed me though, was his use of a diagram of Buckminster Fuller's tensegrity concept. This tensegrity model in the form of a little 3D toy (see diagram on page 42) has been very helpful as a concrete physical example to help people visualize how the suspension system is erected in a web of its own tension. I must emphasize that my model certainly has aspects of a pure tensegrity structure, but differs in an important way (details of this difference are explained later in Part 6 "*More Suspense*").

A decade or so later a colleague pointed me towards Donald Ingber, an American researcher who had come to some of the same conclusions about elastic suspension, albeit just at the microscopic cellular level. (Oddly to me, though, when I met Ingber in person and shared some of the aspects of my whole-person suspension model he did not seem particularly interested in the possibility that the same principle he had discovered on a microscopic level operating in cells might apply to the entire functioning macroscopic being… So it goes.)

For myself, as the years have gone by I have become much clearer about the significance and implications of this model and have seen it proven over and over in practical application with hundreds and hundreds of students. Many of the people I've trained are also out there teaching it to others privately and in universities and conservatories around the world.

Of course, from my current perspective, if I had it to do all over, I'd write a very different set of articles. In fact I am doing just that in a new book. The new book, which is well over half written, will finish the story and bring it up to date…

In the meantime I am excerpting parts of that new book in e-book format. For further information and to sign up to be notified about the publication of that e-book version go to: www.learningmethods.com/revealing.htm.

David Gorman,
Toronto 2012

In Our Own Image

First in a Series on Human Design and Function
Reprinted from *The Alexander Review, Vol. 1 No. 1, Jan. 1986,*
Subsequent editing, June 1993

And Man said, Let us make ourselves
in our own image, after our own likeness:
and let us have command of ourselves;
over all the activities of our bodies,
and all the attentions of our minds,
and over all our creations that creep
upon the earth, and all our imaginings
that have yet to be brought forth.

The General Particulars

This series is dedicated to exploring how we can constructively think about ourselves. To accomplish this we'll have to take into account not only our physical structure *(our design)* but also how that anatomy is designed to work *(our function)* and, most important of all, how we as conscious, energetic creatures go about in the world doing our activities and being aware of ourselves *(our use)*. Our conceptions about ourselves and the way we use ourselves are, of course, the same thing and may or may not be at all in line with the way we're designed to optimally function.

As anyone knows familiar with the Alexander work knows (and especially teachers of the Technique), quite a radical change in our approach to ourselves must be made from our old habits if we are to have any significant improvement. Obviously, along with 'poor use' in any physical sense goes an equally unconstructive habit of thinking. So along with any positive change must come an equally radical shift in our conception of ourselves. I can't reach out of these pages with my hands and deal directly with your physical use at this moment, but we can, with words and ideas, reach into our old habits of thinking (our misconceptions, contradictions, vague assumptions, murky conclusions) and re-organize them into clearer and more consonant patterns in line with the way we're actually built and in line with the way we'd like to function.

I don't propose to go into much technical detail or jargon for the simple reason that I don't want to discuss *THE* Human Body as an anatomical, physiological and mechanical entity as it is usually done. With all its faults this traditional, objective approach has a great deal of value and is a very powerful tool for a doctor or a therapist to use in order to accomplish healing in other people's bodies. However, it is most unconstructive as information that we might want to apply to our own selves from within.

Here these details only seem to confuse and separate us from ourselves. Instead, I am going to approach this with a more personal, less abstract point of view which includes in *YOUR* human body the living, thinking you. If we are to learn to use ourselves well in everyday life we need a way of understanding which uses everyday language and which stays within (or at least keeps returning to) our everyday experience of ourselves.

There are many levels over which we will have to range in order to make sense of ourselves. The most detailed is that of physiological mechanisms like the stretch reflex or how muscle is activated. This is not a level on which we actually perceive ourselves in the normal course of things and hence is of little relevance except in as much as we need to understand some of the implications of these details for the larger view.

Next there is the level of the usual anatomical bits — the named bones, muscles, ligaments, etc. like the femur, the scapula, the finger flexors, the diaphragm. We also do not normally experience in a direct and isolated way these bits and pieces, though we will need to refer to them as the constituents of the next level.

Where we do really begin to experience and use ourselves is at the level of parts (your head, your hand, your fingers, your back, the front of your thigh, etc.), and junctions between parts (your elbow, your hip, your knee, your neck, etc.), and at the level of functional areas (your breathing, your voice, your walking, etc.). This is the common currency of how we feel ourselves and talk about ourselves. But as the common currency of our perceptions it is at this level we find the root of most of our common problems.

This is so not only because to concentrate our attention here is to remain a collection of parts, but also because our attention is naturally drawn to those areas where we have problems generally to the exclusion of other areas. In addition, these elements of experience define quite different territory for different people. For some, the experience of their head will extend downward and include part of their actual neck; some have their breathing taking place in completely different geographical areas than others; and some people perceive their hip joints inches higher up in their pelvis. These sort of mismatches between different people or between an individual's perception of himself and his actual structure have immense repercussions at the next broader level.

At the next higher level we begin to enter a brave new domain. This is the level of interaction *between* our parts and areas — how the use of your arm affects your breathing, the consequences to your walking when you arch your back, how the freedom in your knees is affected by lifting up the arches in your feet — a somewhat intangible level which is not this part or that part but how they get along together. This cooperation or interference can include connections between seemingly distant parts of ourselves — the effect of locking your knees on your voice, for example. It is an understanding of these patterns of interaction that can reduce the mystery of so many of our chronic complaints. We'll spend much time roaming about here.

But there is an even broader and more important level — the level of the organization of these interacting parts and functions into a sensitive, responsive whole. This is, of course, the level of you, the person reading this article. It is this central organization that determines the quality of the interactions spoken of above and the state of health or damage of our parts. And it is the whole person who determines the central organization of themselves by the way they perceive themselves and their activity. That organization can be regarded as our approach to ourselves, our policy of being, our way of working — our *Manner of Use*. That use of ourselves can be one where the way we organize ourselves draws together into a harmonious whole all our parts and functions so that they all work well in themselves and actually augment the function of all the other parts — *Good Use*. Or it can be a way of operating which does not tap into the in-built central harmony of our system so that not only are our parts being interfered with in their own functions,

18

but they in turn interfere with functions around them so that we end up working against ourselves — *Poor Use*.

Thus we come full circle; back to the way we think of ourselves. We will need to dip into the details to help explain the whole, but we'll always need to keep coming back up to that whole, the nature of the central organization — what Alexander called the Primary Control — if we are to gain a deep understanding of the principle behind the Alexander Technique.

Let us now begin with a look at your general physical organization. The most fundamental of your characteristics is your endedness, your extension in space in a length direction. This basic vertebrate quality of *length* is a product of your spine. The spine is the primary organizer of the torso. It is, first and foremost, a lengthening device; hopefully also a *flexible* lengthening device. You are, in a manner of speaking, a flexible torso with appendages.

Furthermore, this endedness of yours is not just an unbiased length, it has a *direction* to its organization. This in-built directedness is toward your head end. Your *head leads* and the rest of your *body follows*. In the embryo, and later as children, development progresses from the head back to rest of the body. Most of your attention to the world is grabbed through the externally-directed senses located in your head; you also search out much information about your surroundings and your own activity by the orienting of these head-located perceptual systems. Most of your intention out into the world is directed out from your head — from your eyes for directing activities, from your eyes, face, and mouth in communication, etc.

In a figurative sense you are like a large arrow with your head being the arrowhead and the rest of you the shaft. A measure of the extent of this head-oriented bias is evident in the feeling many people have that 'they', the point of consciousness, live up in the control booth of their head somewhere behind the TV screens of their eyes whence they operate their body like some large and ungainly machine — there's *me*, but then there's also *my body*. While this feeling is an unfortunate separation of the point of the arrow from the rest, it can be seen to have a basis, however extremely exaggerated, in our inherent nature.

There is a very big difference, though, between the general organization of your torso and that of your limbs. Your arms and your legs are much more similar than they are different. They have the same basic patterns of bony structure and muscular arrangement and are so built that most of the bones are long and stick-like. The bones run down the centre of the limb with the soft tissues arranged around the outside. These soft tissues are not organs as we usually think of organs, but mostly muscles, tendons, and ligaments having to do with the operating of the limb. They tend (the muscles in particular) to be long worm-like in shape and run more or less along, or curl around, in the direction of the length of the limb.

Within those similarities, your arms and legs differ in virtually opposite ways. Your arms take their rooting and grounding from your torso and extend out into the world embodying a large number of possible movements that we can put together in various combinations. In the most general sense, your arm is a means of getting your hand around so that you can manipulate things and make changes in the world. Your legs, however, take their grounding and support from the ground. They operate, for the most part, up at their other end to support and move you around. In this same general sense, we could speak of

19

arms as having an *active*, explorative nature since they are normally used to do things and explore your surroundings. We could speak of legs as being *responsive* since their activity is usually a response to the activity of the whole of you.

Your torso has quite a different organization reflecting its older origins, evolutionarily speaking. Here the bones are arranged around the outside enclosing a space inside for your organs. The musculature here tends to be broad sheets in overlapping layers which run between the bones to complete the walls of this space. Your torso, in other words, is primarily a volume, a *container*, for your life support organ systems. While your limbs tend to be about activity and *doing* (to be even more general) your torso tends to be about *being*, with your organs providing the ongoing energy and metabolic support for your doing.

Let us look more closely. Your head is clearly a bony container for your delicate brain, as well as a framework for your major external senses and the openings to your digestive and respiratory systems. The ribs in your chest have the primary job, not of protecting your organs, but of making sure there is maintained the important volume for your heart and lungs (even though the whole point is that it be a changeable volume so that free breathing can take place). Down in your abdomen the container has mostly muscular walls for the quite variable volume of your digestive system. And finally your pelvis forms a strong and firm base for the container as well as a framework for your reproductive organs and the outlets of your digestive and urinary systems.

Connecting all these areas is your spine, the major lengthening device of this container. Without your spine your head would be inside your ribs, which would be sitting on top of each other, all of which would be piled neatly inside your pelvis. Your spine, in other words, is a *spacer* that takes care of the expansion of the container in a length direction. From this length the other parts take care of an expansion outwards into width. The length is primary; without it there can be no effective width, as we shall see in future articles.

Figuratively speaking, your spine is the limb for your torso. Those prongs you feel sticking backwards from each vertebra may be right under the skin of your back, but the actual limb-like column of bones runs up inside the centre of you at least as much at it runs up your back. In fact, in your neck the spine is virtually in the centre with your throat in the front half and masses of muscles in the back. It deviates to about three-quarters the way back through your chest then comes forward in your lower back to about two-thirds the way back. The spinal muscles, too, arranged along the length of the spine tend to be somewhat long and limb-like. Indeed, due to evolutionary changes, your neck is not so much an organ container now as it is a limb for your head. It's your head's way of getting around.

If your head leads and the rest of your body follows as an in-built organization, and your spine connects your head to all your other parts, then there are obviously some very important interactions that take place between your spine and these other areas, particularly up at the head end. In future articles in this series I'll look into these interactions and continue to throw light on this inherent Primary Control. Next issue we'll explore the repercussions of the completely different kind of joints in your torso compared with your limbs and what constitutes the optimal environment for your life-support systems.

In Our Own Image

Second in a Series on Human Design and Function
Reprinted from *The Alexander Review, Vol. 1 No. 2, May 1986,*
Subsequent editing, June 1993

The Nature of the Torso

Last issue we began to look at your general physical organization; how you have an endedness, a length; and how that length has a bias to it, a direction towards your head end. We looked at how that length is embodied in your spinal column and saw also how your limbs (in a very general sense) have a structural organization for *doing*, whereas your torso has a general organization of simple *being* as a container for your life-support organs.

Let us now proceed to probe further the nature of this container. There is more to the difference in character between your limbs and your torso than just the structural arrangement of the bones and connective tissues (muscles, tendons, ligaments, fascia, etc.). The limbs and the torso also have a major functional difference in the way their parts are connected and move about on each other. That is, they have very different kinds of joints.

When we think of movement and joints, the limb type of joints are the most familiar to us: the elbow and hip joint, the finger joints and so on. While these differ in shape and range of motion they are all the same basic kind of joints — synovial joints. The joint surfaces where the bones meet each other are covered in smooth cartilage and the entire joint is surrounded by a fibrous capsule lined with a membrane (the synovial membrane) which secretes a very slippery lubricating fluid. In these sort of joints, bones move with little resistance on each other to take up new relationships. In consideration of how we normally think of them, I call these simply *moveable joints* because they allow *free movement* much like a door moves on its hinges throughout its range and can easily stop anywhere within that range.

Torso movement is not at all like this. However much your torso is a container it is obviously not rigid like a tin can (unless you hold yourself stiff). You are designed to be able to bend and turn while at the same time allowing your organ systems to carry on their functions inside. To say that we are capable of this, however, does not mean that we are always able to do it. To understand the joints which allow this torso flexibility let us first delve into the organization of the parts concerned. These parts include the central length of your spine, the strong pelvic bottom of the container from which your spine arises, the volume created by your ribs midway up your spine, and your head containing your brain and framing your major senses.

The bony vertebrae that make up your spine are separated from each other by thick intervertebral discs. These discs are made of a compliant and elastic cartilage reinforced with connective tissue fibres and bonded very firmly to the vertebra above and below. Your ribs are connected together in front to the bones making up your breastbone by bars of pliable cartilage resulting in a flexible chest volume. The two halves of your pelvis are joined together in the front at your pubic bones by a disc of cartilage. At the back they are joined to the base of your spine on either side of your sacrum by what are technically synovial joints but which are so strongly bound by ligaments as to not be moveable in the same sense that your fingers are moveable. At best the slight give and take possible in these sacroiliac joints could be called *resilience*. A similar situation of limited but elastic movement is present in the three bones making up your sternum (breastbone). The dozens of bones fitted together with suture joints in your cranium also partake of this resilience though on an even smaller scale that that of the pelvis and sternum.

Thus your torso is primarily a distortion structure. There is no movement of your torso, be it turning, bending or straightening, without the discs in your spine being elastically distorted in some fashion. You can bend over forwards because each one of the discs involved in the movement can be squashed in front and stretched out behind to allow the bend. Most movements of your torso will also involve the resilience of your chest and likely, to a smaller degree, a give and take in your pelvis. Thus, these torso joints I call *distortion joints* (with no particular negative connotations to 'distortion'). They give you a kind of movement that might best be termed *flexibility*. They have the pliancy to allow you to bend and turn like a young willow branch, but as you do so, elastic resistance to distortion increases and they always want to recover their shape and return to where there is as little distortion as possible.

In your torso this inherent shape is where there is the maximum volume for the container. In terms of the spine it is where the discs are neither squashed nor stretched, neither twisted nor bent; where the spine is in its natural curves and at its natural length, neither over-bent nor over-straightened.

A glance at the spinal column shows that it curves back and forth. These curves are not there because it is like a strand of limp spaghetti buckled under the weight of years; they are built into the shape of the bones and the discs themselves. This does not mean, though, that whatever curves are in your spine will necessarily show up as corresponding outwardly visible curves of your back. A simple exploration will show this well. Touch your spine in your lower back. Note that the columns of muscle on either side bulge out farther than the prongs in the middle. Those prongs are just the tips of bony projections from the vertebrae; the actual bodies that form the lengthening device are at least two inches deeper inside your back. But farther down, above your tailbone, the bony base of the spine (your sacrum) is right under the surface, and farther up between your shoulder blades the vertebral bodies are only about one inch from the tips of the prongs.

You can appreciate from this that if someone is sitting and their lower back looks somewhat straight you are seeing only the profile of the muscles covering quite a substantial curvature of the bones inside. I have had just as many students tell me they have arched their backs because they've been told their spine is 'supposed' to be curved as I've had students who tried to force their curves straight because they think of that as good posture. In either case they have only been able to get themselves 'right' by what they look like in a mirror which, unfortunately, reflects back not at all what is going on under the surface.

The whole point of this is that the spine is a curved structure as well as a distortion structure. The curves are important to give strength to the spine and resistance to deformation; they provide a springiness and a shock absorbing effect simply not possible were the spine straight. But most of all the curves provide a resolution of any conflict between the need for a lengthening device and for volume of the container. In your neck the spine moves forward to be as central as possible under your head. There are not many sizeable organs in your neck and the need for volume has become secondary to providing mobility and range for your head. There is even an elaborate arrangement of synovial joints between your head and the upper two vertebrae, the significance of which we will explore in future issues. Your spine moves back in your chest to allow room for your heart and lungs then moves forward again in your abdomen to be more central for the mobility of that area. Finally it moves sharply back to fit into the pelvic bottom of the container and to be out of the way of the outlets of your digestive and urinary systems and reproductive system (particularly for giving birth).

This leads us back to the reason for the distortion joints. The torso is first and foremost a volume for your organs. These organ systems — your respiratory and circulatory systems, digestive and excretory system, — merrily breathe, circulate, digest and excrete away without us having to do much directly to help them. They are not an activity to be done but a response to the activities we do. These systems work together to provide the energy and vitality for your activity. The best you can do for them is to ensure that they have the optimal environment for their various functions and then let them get on with it. That is to say, your job is to find how you interfere and get out of the way.

This optimal environment has two major characteristics: lots of room *and* lots of freedom. The bony arrangement of your torso is designed to afford the room; the distortion joints are designed to afford the flexibility. They will let you move about but always want to return to their inherent shape — where you have the maximum length and the maximum consequent width *with* maximum freedom. Now this doesn't mean that you should always stay close to home and never bend; the mobility is there for you to use freely. If you use yourself freely and openly to go off into movement you will be more free and open wherever you go and you can certainly use yourself freely and openly to come back. The pure fact of it is that free bending and turning is a response of your torso to your activities and will occur naturally unless you stop it by stiffening yourself.

As an aside here… the major problem with excessive torso movement arises when we don't have the *skill* to use our large synovial limb joints (hips, knees, ankles, shoulders, and so on) openly and so force ourselves to compensate by using our distortion torso joints for active movement. Skill is the operative word here because the torso was the original and entire body of the early fish vertebrates before they put on legs and came out of the water. Distortion movement is much simpler and more primitive than that of the sophisticated synovial joints (which have been proportionately increasing as vertebrates have progressed). If we do not claim our heritage by gaining command of the range and freedom of these synovial joints (including the ones between your head and the upper vertebrae!) we have no option left but to shut them down by locking or fixing them with muscles and to fall back by default on simpler distortion movement which is neither designed to nor capable of doing the same job as the freely moveable joints.

In any case, all your organ systems must still go on working for you even if you restrict their space. They simply go on working under pressure. Each of your organs has a specific form inseparable from its function. There are no empty spaces where organs can go when you squash them. Your heart has to pump just as much blood, it simply has to pump thousands of times a day that much harder, and since it has to work from a cramped and distorted space, it will pump that much less efficiently. It may take years, but sooner or later this will lead to a breakdown in function.

However, it is not so much your organs that directly bear the brunt of a slouching squash or a straightening squeeze, it is the flow *between* your organs. Your life ultimately depends on the flow of air in and out of your lungs; the flow of blood carrying oxygen around your body and picking up wastes, the flow of food through your digestive system so nutrients can be absorbed, and so on. If there is ongoing distortion and pressure you are sitting, not at all figuratively, on your metabolism and squeezing the life out of yourself.

It is easy to appreciate how constriction and compression of the container can adversely affect the health of your organs and your state of vitality. It is not always so easily appreciated how lack of flexibility is its accomplice. Respiration is the best example. Breathing is the most noticeable rhythm in your body. It is literally the breath of life; its freedom and depth the most direct measure of the liveliness and energy available to you. Unfortunately (in some ways), it is also the one of your life-support organ systems with which you can most directly interfere, since it must make use of much of your voluntary musculoskeletal system. Its major requisites are just the same as those of your torso as a whole — a volume that is free to move.

One of the most common of interfering habits, right up there with bad posture, is good posture. Bad posture most often means some version of slouching or collapsing forward with obvious consequences of cramping and jamming up your breathing. Good posture is usually the opposite, with attempts being made to straighten up and lift your chest, pull your stomach in and shoulders back and all that. The way most people get their chest up in the front is to pull down on their back with those powerful muscles that run up either side of the spine. These muscles do not just attach to the vertebrae, they also spread out to attach several inches onto the back of each of your ribs. Thus any pulling down to raise up your front and hold it, pulls down on your ribs and *holds* them down losing some volume and freedom. Since your ribs slant downwards from your spine in the back, not only are they pulled down but they are also pulled in towards the spine, narrowing your chest losing more volume *and* freedom…

If this isn't enough, any thoughts about your abdominal muscles being strong in order to hold your guts in only add to the problem. These muscles all pass between your pelvis below and your chest above. So when you try to flatten your stomach and narrow your waist you are shortening these muscles strongly and tending to pull your chest to your pelvis. Of course the pull down in your back is then partly to hold up the pull down your front so you don't bend over forward. But now you have two downward pulls and breathing is even less free. This is not the end of it. Of the abdominal muscles, only the one in the front actually goes up and down between your ribs and pelvis. The others run obliquely and crisscross from the ribs on one side down towards your pelvis on the other so that their tightening not only pulls your ribs down but also pulls them together in the front — all of this the opposite of free, open breathing.

I could go on, bringing in how tension in your shoulders clamps your breathing in a vice or how holding your hips joints stops the flexible breathing rhythm dead at the bottom of your torso, but the point is clear. It is the *freedom* of the parts that allows free breathing and free breathing is the opening of your torso into volume. Free breathing is also openness to what's around you. It is you letting the outside into you and allowing what's in you to go out. It is animation and activity, expression and response.

The openness and expansiveness of your torso as a whole vitalizes you and liberates your energy so that it rises up in you. And that, more than anything else, is what gets you up and keeps you up, not bones and muscles, or balance and posture. Next issue we'll begin to look at how that energy keeps us up.

In Our Own Image

Third in a Series on Human Design and Function
Reprinted from *The Alexander Review, Vol. 1 No. 3, Sept. 1986,*
Subsequent editing, June 1993

From the Ground Up

In the previous instalment we looked at the distinction between the freely moveable joints of the limbs and the distortion joints of the torso. We saw how these distortion joints provide flexibility to the container of your torso as well as allowing the rhythmic change of volume in breathing. If we can develop the skill to allow this flexibility and freedom as well as an openness and volume we ensure that our life-support organ systems have the optimal environment to provide us with the vitality we need to be active and alert.

In this issue I wish to explore how deeply linked this optimal environment is to our characteristic human uprightness — indeed, how dependent our whole freedom, openness and awareness are upon our skill of orienting ourselves to the Earth and the sky. While it is perhaps somewhat of a dubious proposition these days for us humans to think of ourselves as 'the crown of creation' there is at least one aspect of evolution where we have gone about as far as it is possible to go: the development of uprightness.

The earliest vertebrates, the fish, evolved under the surface of the planet and moved horizontally in the water following the direction of their length. The new creatures who emerged from them gradually but unceasingly streamed upward in the world. As life progressed, animals came out onto the surface of the land, long and low with wide-spread legs. From these in turn sprang beasts with their bodies well up off the ground, legs well in underneath them and with longer necks so their heads were raised high up. Some of these even came up off their forelegs to become bipeds though their bodies were still angled forwards with long tails balancing them behind. This tendency to raise up parts of the body over other parts, higher and higher above the ground, led eventually to the primates and us with our long straight legs, our opened-out hips and erect upper body.

From the water in which we evolved we have raised ourselves a full 90 degrees to be stretching up from the earth toward the stars. At the same time we have come a complete circle back to where our whole body extends back from our head just like the fish. It is difficult to conceive what other of our parts could be moved up higher short of growing antennae from the tops of our heads.

Such a sweeping trend toward height would not have been so consistently pursued were there not huge advantages. One of the most obvious is that as the creature's head is raised so are its senses and thus it comes to live in a larger world with larger horizons. It can smell more, see more and hear more. A bipedal creature has the additional bonus of freeing up its front limbs for other functions. A less apparent but much more important benefit of having mass poised up high is the gain in potential energy and hence the ease of initiating and augmenting movement. With this there is however a penalty of sorts to pay. Nothing is so easy or so stable as resting on your belly on the ground. The higher up that animals became the more unstable they became and hence the more complex strategy needed to maintain themselves up there during activity.

In our uprightness we are probably the most precarious of all the creatures with so much of our body so high up over so many free joints. Precisely because we are so much up over ourselves, we also have the possibility of being one of the most balanced of creatures (albeit a possibility more often than not unrealized). In fact, it is this very instability in our balance that enables us to effectively use the potential energy of our height. At the same time it is the balance in our instability that allows us such a potentially low overhead in terms of energy expended in staying upright.

Let us look into this a little more closely. Firstly, to separate somewhat the inseparable, in the last issue I distinguished between *doing* and *being* as respective characteristics of the limbs and the torso. Here I wish to broaden this to distinguish between *doing* in the sense of intended activities and movement, and *being* in the sense of simply being a creature, living, breathing, aware, and upright, but not necessarily engaged in some actual visible endeavour. In this context your ability to *be* upright and relatively free is a prerequisite to *doing* anything you may want to do. Not much can be accomplished in the daily round if you keep falling over. We can go further — the freedom, ease and coordination of movement utterly depend upon your primary skill at maintaining a free, poised and open uprightness. If you can't even stand up or sit up without tightness or collapse, you haven't a hope at ever *moving* freely. If you can come to a well-supported flexibility and an openly integrated being in any moment in yourself, you have the possibility of taking this free being off into activity.

The whole issue of how we use ourselves in relation to gravity is normally termed *posture*. It is clear that a large part of any success at being free rests on our ability to use well the support of the ground so as not to be increasing our instability by leaning one way or the other. This is usually taken into account with the concept that 'good' posture has something to do with *alignment*, which in turn has connotations of vertical and straight. There is much truth in this from the point of view of simple physics, however, there is nothing simple about it. To appreciate the full implications of our relationship with the earth it is necessary to go back to our actual given structure and ask how possible is it for humans to be balanced in the sense of vertically aligned up over the ground? And furthermore, how possible is it for humans to so balanced up over their major joints?

It is certainly possible to achieve this balance between your whole body and the ground. In fact, you are conveniently built so that when you have the planet directly underneath you there is an even distribution of contact on the bottoms of your feet — as much to the front as the heel, as much to one side as the other. There you are totally supported. When you have the planet directly under you there is no way on Earth that you can fall. After all, you are already on the ground and there is nowhere lower to fall to than the ground. Even more conveniently, the bottoms of your feet are extremely sensitive to support provided you are attentive to and able to interpret the information they tell you.

And very important information it is too. All this may seem very elementary, yet surprisingly, so many people consistently carry themselves with more contact on the front of their feet (which means they are leaning forward), or more back on their heels (leaning backward), or even habitually stand more on one leg than the other (with consequent twists all up their body) and then wonder why they have these tensions and pressures which they can't release. They can't get rid of them for the simple reason that the tensions and pressures are all that is holding them up. Were they to succeed in releasing these habits without changing their orientation to the planet, they would simply fall over.

Even more surprising is how many people go into movement by going even farther out of support. They begin to walk by leaning forward or to sit down in a chair by going off balance backwards towards the chair and then tightening and struggling with the imbalance all the way to the chair. More often than not they never regain their balance and would be in big trouble if there was no chair. It is no wonder that many people feel awkward and out of control in themselves (a feeling guaranteed to give one the impression of being separate from one's body and hence rather more refined a spirit than this contrary vehicle in which one finds oneself).

I do not wish to make too much of an issue of this but we too often devote a lot of awareness to what is going on *inside* our bodies and pay selective inattention to what is going on between our body and the planet — to how well the whole of us is supported. That sensation of contact on the bottom of your feet (or under your seat and thighs if sitting) is your direct awareness of the earth pushing up underneath you and supporting you. It is your tangible appreciation of your relationship to the planet and to gravity. It's very direct and very reliable, it's there any time you wish to be sensitive to it, and it is very easy to respond appropriately to this sensitivity. The more even and broad is your distribution of contact with the ground (with consideration for the circumstances of your activity), the more you are well supported and the more you have a chance of being free. Anything else is just unskilful use. We'll see just how important this is as we begin to uncover how curious a beast freedom can be.

What we have been looking into is not only our ability to balance ourselves on the ground but also to poise ourselves on our feet at our ankle joints. When we look at the uprightness of the rest of our structure we find that very different instabilities exist in our legs than in our torso and, due to the difference between the freely moveable synovial joints of the legs and the flexible distortion joints of the torso, very different strategies of balancing are required.

Since your hip, knee and ankle are huge synovial joints with large ranges of movement it is conceivably possible to balance yourself on your legs by getting your torso exactly up over your hip joints, the rest of you just balanced on your knees, and the whole of you poised over your feet (which are evenly supported on the ground). But these are freely moveable joints and will not stay in that poise unless you can stop your breathing and your heartbeat, neither talk nor move your arms about, and, of course, pray that no wind comes your way. The only way you can stay in such a 'balance' is to regard it as a 'position' and then hold the joints with muscles so that they are immobile and can not be disturbed. There is then, however, no freedom. Thus, in your legs you cannot have a balance as we normally think of it and still have freedom of the joints. But you can have a 'balancing' — an equilibrium — that is gained, lost, and continually restored. In short, a *dynamic equilibrium* and one which hopefully is gained more or less completely, which is lost only a little, and which is restored often.

The situation is not at all the same in your torso. Here, it is not possible to balance in the sense of getting all the parts up over the ones below. We just happen to be built so that we are *inherently unstable*. A glance at a skeleton from the side clearly shows that there is more of you in front of your spine than behind. When you think of it, this is quite obvious from experience. When tired you tend to sag and slouch out forward. If you slip off to sleep while sitting up your head nods out to the front. Thus, your head and torso are inherently unstable in a forward direction. In addition, we commonly bend forward in activities or use our arms out in front of us to do something thereby actively de-stabilizing ourselves further forward.

This forward instability is a fact of our structure which we can do nothing to change but, obviously, something must be done to deal with the instability if we are not to be collapsing over all the time. Something must also be done so that any instability above does not overwhelm any poise achieved in your legs below. The traditional solution, as contained in the concepts of good posture and alignment, has been to *straighten up*, which is to say, to attempt to reduce the instability by doing work to raise each part more up over the parts below. Successful alignment — good posture — is presumably where sufficient stacking up of the parts has been achieved so that you are 'centred' and the amount of effort in holding yourself up can be reduced. You would then be in a state of 'ease' and 'balance'.

There is, however, a major flaw in this system of approach. The major joints of your forwardly unstable torso — the vertebral disks, the rib cartilages, etc. — are not synovial joints which would allow your torso to be easily moved to a new, more vertical position and to freely stay there. They are cartilage distortion joints and while they have a flexibility that allows movement, this very flexion is accompanied by an elastic resistance which increases with distance resulting in a strong tendency to spring back to their previous shape. This means that not only do we have to use muscular work to raise ourselves back up out of our forward instability, we will also have use muscular effort to hold ourselves up there against the elasticity of the distortion joints. This is why 'good posture' is so often hard work and leads to stiffness and tension. It is also why the consequent release of tension and holding when we 'relax' usually means a blessed collapse. The upshot of this approach is that in the attempt to achieve an upright freedom it is all too easy to find oneself stuck between being 'upright' but not very free or free but not very 'upright'.

If we track it down, we find that the flaw exists in the assumption that instability is undesirable; that there is something wrong with the way we are built — perhaps an evolutionary oversight to be corrected in future generations. It is often said in the books that we came upright so quickly (in an evolutionary time frame) that parts of us were left behind. In particular, they conclude the lower back and pelvis area is in evolutionary arrears which is why humans have so many chronic problems in that region. However, the flaw is not in our design so much as in our function, or to be more precise, in our malfunction which, in turn, is the product of a rather simplistic and mechanistic conception of instability.

There are higher forms of stability than a simple and static balance of objects on top of each other where any disturbance leads to a loss of balance and must be resisted. In these higher systems instability is not only an integral part of an organism but is the driving force leading to a larger stability of the organism. In the next several issues we shall see how this applies to us (and the seeming conflict of being simultaneously up and free) by turning around the premise that our design is a problem to be dealt with. We shall see where it leads if we assume instead that our inherent instability is there for a reason and that to reduce it is to lose the very freedom and liveliness we are after.

In Our Own Image

Fourth in a Series on Human Design and Function
Reprinted from *The Alexander Review, Vol. 2 No. 1, Jan. 1987,*
Subsequent editing, June 1993

Talented Tissues

The previous issue of this series finished with a promise to explore how our inherent instability is not a design fault and thus a postural problem to be solved, but rather an essential part of our freedom and responsiveness. In order to proceed, our first step must be to understand the nature of the system called upon by this instability, the *connective tissue* system, and then to understand how the resulting responses, in turn, raise the whole being to a higher stability.

The connective tissue system, for our purposes, encompasses not only the set of passive tissues normally called "connective tissue" (ligaments, tendons, fasciae, etc.) but also the active muscular system which is an evolutionarily specialized form of connective tissue. These are the tissues which keep us together, or more appropriately put: these are the tissues which keep us from coming apart. For it is when there are disconnecting or de-stabilizing forces that these tissues are brought into play. That is to say, their primary job, collectively, is to respond to disconnection by resisting it. Or to put it another way, these disconnecting forces draw upon the tissue's length and their response is to match these forces with the integrity of that length.

Regardless of what name or what form it takes, passive connective tissue (as opposed to more active muscle), is all composed of the same material — fibres of *collagen* and *reticulin*, white and relatively inelastic, and *elastin*, yellow and very elastic. The various forms of this tissue are different simply by virtue of their location, their function and how the fibres are mixed and arranged within the tissue, be they tendons from muscles to their attachments, ligaments interconnecting bones near joints, sheets of fascia surrounding groups of organs or muscles, aponeuroses like flat tendons spreading out to or from a broad attachment, or the myriad of irregular fibres which interweave through the skin or just run individually from here to there connecting everything to everything else.

Before going into the characteristics of connective tissue it is worth clarifying the intimate structural relationship between "passive" connective tissue and "active" muscle. Muscle is a highly organized tissue with many levels of structure. Each distinct, named muscle (e.g. the biceps) is composed of thousands of long and exceedingly thin *muscle fibres*, individual muscle cells which are about one quarter the width of a human hair and may be up to a foot long. These can run in various directions in the muscle but usually run more or less in the direction of the length of the muscle, that is, more or less in the direction in which the muscle is called upon to respond to forces.

Each of these muscle fibres is contained and surrounded by a connective tissue sheath to which the complex inner, contractile, protein structure of the fibre is attached along its length. The muscle fibres with their investing sheaths are gathered into bundles called *fasciculi* by means of another connective tissue sheath and finally these bundles are bundled together by another sheath around the muscle as a whole. These sheaths, at all their different levels, carry along the length of the muscle to the ends where they thicken and fuse together to form the tendon and ultimately insert to its attachment. Thus the tendon is not just stuck onto the end of the muscle and attached to a bone, but runs right through the muscle and out the other end.

Bearing this in mind, we can differentiate between the active part of the muscle, the red-meat, protein structure inside the muscle fibre which responds electrochemically in concert with the nervous system, and the connective tissue part of the muscle which forms a considerable part of the bulk of the muscle and a very important functional relationship in the muscle as we shall see.

The passive connective tissue is arranged in the body in precise opposition to disconnecting stresses and its major characteristic is that it is constantly trying to shrink, that is, to shorten in itself against these disconnecting stresses. It will shrink up in this manner until it becomes under enough tension to prevent it shortening any further. The resistance to the shrinking which brings the connective tissue under tension is, of course, from us going about our daily activities, bending, turning, flinging our arms about and constantly subjecting the tissue to lengthenings and stretches. The ongoing result being that the connective tissue will always shrink up to the edges of your range of motion. For an individual so unlucky as to have had an accident such that they are in hospital in a coma this shortening of the connective tissue will become so marked as to tighten them virtually into a foetal position. To keep the joints open and free physiotherapists must supply the range of motion by regularly moving the patient.

Similarly, if your daily habit of use were to become more tight and narrow, the connective tissue would gradually take up whatever resulting slack by physiologically shortening until it comes under stretch from your new, more restricted range of movement. If you subsequently were to release these compressions and use yourself more openly, the connective tissue would be brought under greater stretch and would gradually yield to become longer tissue. All this happens bit by bit over a time scale which would produce noticeable changes in a period of several weeks. Thus, I've come to call this connective tissue passive, since it is not directly responsive to your thought in the manner of the active

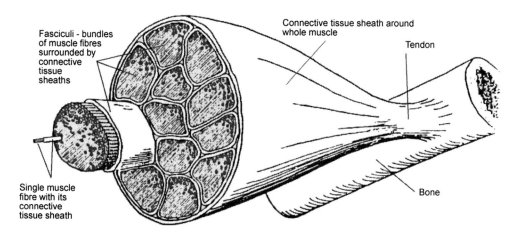

Figure 4.1. The Organization of Muscle (not to scale)

muscles. Instead it follows your activities, adapting to them and defining your limits at any one moment as a product of the use of all previous moments.

We often tend to think of and feel our connective tissue as restriction which prevents us from having a broader range of motion and thus begs for stretching exercises to open it out giving us more "room" to move. But it is important to be aware that the connective tissue is not there to get in your way, rather it is your basic level of integrity and security. The point being that it doesn't matter what you do, the connective tissue will always tighten up to the borders of that use, giving you the range within which you have the skill and practice to operate, but securing you from any precipitate and potentially damaging excursions beyond.

Since most of our movements take place comfortably within that normal range, the occasional activity which ventures outside is bound to meet the resistance of the connective tissue. Any stretching and opening of that resistance will be transitory if the majority of time is spent operating well within the familiar range. Expansion will only be able to be maintained when the consistent day-to-day use also expands to use this new space fully enough to keep it open.

The larger picture of how connective tissue contributes to the higher stability of the individual is inseparable from its interactions with muscle, the active part of the system. Muscles are not only highly organized but also very talented structures. We have been primed by our normal focus on doing things and achieving ends to think of muscles as having to do mostly with movement and activity. But by far their most primary function is to maintain our integrity and security as upright creatures against all the instabilities and forces tending to disturb our equilibrium. The ability of muscles to be engaged in movement is a secondary function only coming into play when we've managed to succeed at uprightness.

Whereas the passive connective tissue adapts over the course of weeks to changes in use, at any one moment it cannot modify itself and has only its current length to resist disconnecting forces. Muscle, like passive connective tissue, also becomes physiologically modified to our changing use over the course of time. However, muscle is a sensitive and responsive tissue. What it is sensitive to is stretch or an increase in tension and how it responds is to generate what activity is necessary to resist that increase, to keep the parts

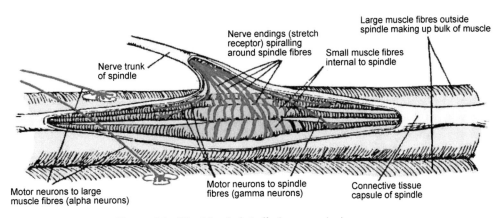

Figure 4.2. The Muscle Spindle (cutaway view)

31

from coming apart. To use our previous terminology, muscles are sensitive to forces of disconnection and respond so as resist that disconnection.

The sensitivity of the muscle is embodied in the muscle spindles, and its responsiveness is a product of a larger sensory/motor feedback, the stretch reflex which I shall now describe (in brief, since in these pages justice cannot be done to its exceeding complexity). The muscle spindles are numerous small structures (about one third inch long) sprinkled throughout the matrix of the length and depth of the muscle. They are made up of a few (6-10) small muscle fibres attached at both ends to the sheaths of the larger fibres which make up the operative bulk of the muscle. Wrapped around the middle of these spindle fibres like the tendrils of a honeysuckle vine is the primary sensory nerve ending of the spindle, the stretch-receptor.

This primary receptor generates impulses into the central nervous system when the middle part of the spindle is put on a stretch (when the middle part increases in length due to an increase in the tension applied to it). It is this stretch-receptor response which initiates the stretch reflex. It is like other reflexes wherein a particular external change stimulates sense organs which send impulses to the central nervous system resulting in impulses being sent out to some other organs, such as muscles, which produce a recognizable response to the original stimulus. In this case it is the stimulus of a pull on the muscle which excites the stretch-receptor sending impulses into the spinal cord from whence impulses along the motor nerves are sent back to the muscle causing it to be activated, to "tone up", to a degree sufficient to just resist or oppose the original pull.

However, it is not just the muscle pulled upon which is thus activated, other muscles around it which may normally cooperate with the stretched muscle are also activated in order to help share the load. What all this amounts to is a feedback loop to automatically adjust the activity of the muscles so as to deal with the stretching (disconnecting) loads upon them. (Actually it is a more complex error-activated feed-forward system, but for our purposes to stick in a simple manner to the main point will not do us any harm.) The beauty of this arrangement is that the amount of muscle activity generated is just enough to deal with the load and no more. If the load should change, so the response will automatically change.

The whole mechanism is not just a simple closed loop where any stretch leads to a consequent resistance or we'd never be able to move. We need something in this loop which enables the system to turn on and off the reflex, or more precisely, to raise and lower the threshold at which it is activated. To make a long story short, that something which controls the reflex loop is YOU — your conscious intention.

This is brought about by mediation of the loop at the spinal cord level and also by the activation of the small muscle fibres in the spindles themselves so as to modify the sensitivity of the receptor. The net effect is that the stretch-receptor is sensitized by your intention of the moment so as to detect any disconnecting or de-stabilizing change as would deviate from your intent and thence to initiate the reflex loop which is completed with the excitation in the relevant muscles of a response to resist the disconnection or instability.

Thus, there are two factors essential for the manifestation of the stretch reflex: your intention and some stretch or pull on the muscles which would tend to deviate from that intent. The stretch reflex is the mechanism down at the muscular level which takes care of a lot of the detail of carrying out your intentions, automatically and reflexly, so that

32

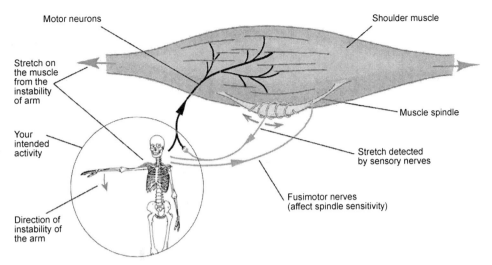

Figure 4.3. The Stretch Receptor

Labels in figure: Motor neurons; Shoulder muscle; Stretch on the muscle from the instability of arm; Muscle spindle; Your intended activity; Stretch detected by sensory nerves; Fusimotor nerves (affect spindle sensitivity); Direction of instability of the arm

you, the conscious creature, do not need to. When you say to yourself, "I want to stand up from this chair", much of the movement and the support and stability throughout the movement is a product of the stretch reflex. Obviously, the thought of "how" you are going to stand up from that chair is the operative part in sensitizing the reflex. However you think of the standing becomes the template to which the detector compares the action.

It's essential to realize (and we'll explore this in more depth as we go along) that this "how" I stand up is not so much a characteristic of the standing up as it is of the "I" who is doing the standing up. What sort of an "I" is standing up? Is it a tight, fixed, breath-holding, pulled down sort of "I"? In which case the standing will have permission to be a tight, breath-holding, pulled-down sort of standing. Is it a poised, freely-breathing, light and open sort of "I"? Then so will be the movement! Let us return from the level of cellular physiology to the larger scale of the whole body. Now we can better appreciate the significance of why muscles are arranged in our body where they are. Muscles in their very nature are built to be sensitive to stretch and to respond by resisting that stretch, so we would expect to find them situated where they would be likely, in the normal course of things, to encounter stretches. If you remember from the previous issue, we are built so that we are inherently unstable at all our major joints. Since our operating uprightness is based upon these instabilities not being allowed to increase and manifest as falling, the support system is arranged with muscular tissue opposite to the instabilities such that it would be stretched were falling to occur.

Your head is inherently unstable forwards as are the other parts of your torso. It is for this reason that you have the massive set of erector spinae muscles extending up your back on either side of the spine from the pelvis right to the skull. The forward instability of the head and torso puts a stretch up the length of these muscles and as long as your intent is to stay more or less upright, the stretch reflex will see to it that these muscles resist that stretch.

The situation is different at your hip joints. Since you are a bilaterally symmetrical creature there is not a large instability side-to-side if you are on both feet. But unlike above,

33

there is instability in both forward and backward directions. The large gluteus maximus mass of muscles between your sacrum/pelvis and the back of the thigh-bone responds to instability of your torso forward on your legs, while the iliopsoas group extending down from the sides of all the lower back vertebrae and the broad inside surface of the pelvis to the lesser trochanter on the inside of the thigh-bone deals with resisting backward instabilities.

Nevertheless, in some ways of standing, and especially in locomotion when one leg is off the ground, your torso is also unstable in a side-to-side direction. Here the other gluteal muscles, gluteus medius and gluteus minimus from the broad outside surface of the pelvis to the greater trochanter on the top of the thigh-bone are activated by instability to the side of the off-ground leg, while the adductor mass of muscles from the pubic bones and bottom of the pelvis to the length of the inside of the thigh are resistant to instabilities in the other direction.

At your knees strong connective tissue ligaments preclude a toppling forwards off your knees and also prevent sideways or lateral movement, hence the major instability is backwards such that the knee flexes. In addition, your own activities of raising your arms

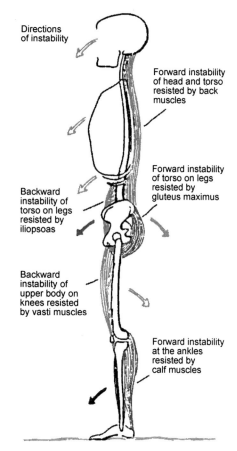

Figure 4.4. *Muscle masses in relation to the innate instabilities of uprightness*

in front of you or bending over further de-stabilize the knee into flexion. The quadriceps group of muscles in the front of the thigh which inserts into the kneecap and then to the tibia in the lower leg is situated to deal with these instabilities backwards off your knees.

When you achieve a reasonably well-supported uprightness the ankle joint is unstable both forwards and backwards. The muscles from the front of the lower leg to the foot, primarily tibialis anterior but also the toe extensor muscles, deal with backwards instabilities. However, due again to your own forward de-stabilizing in reaching, bending, and locomotion, there is the much larger calf muscle mass in behind the knee and the back of the lower leg running down to the heel.

These major muscle masses — the back muscles, the gluteus, the quadriceps and the calf — have developed to respond to our unstable uprightness thereby giving us our characteristic human muscular shape. We shall continue in the next few parts of this series with how these are best seen not as separately-acting discrete muscles, but as part of a whole system including every muscle in the body, a flow of muscle responsiveness, which can not only support us in relation to the planet but can do so without in any way restricting our freedom of movement.

In Our Own Image

Fifth in a Series on Human Design and Function
Reprinted from *The Alexander Review, Vol. 2 No. 2, May 1987,*
Subsequent editing, June 1993

The Suspension System

Two issues ago we looked at the inherent instability of our structure in relation to the planet. Last issue we went on to the nature of the connective tissue system which deals with this instability. In particular, we looked at the primary talent of the active muscular tissues — being sensitive to stretch (disconnecting forces) by means of the stretch receptor and via the stretch reflex being responsive to these forces by resisting them.

The activation of the stretch reflex, if you remember, requires the satisfaction of two conditions: your *intention* which acts as the general policy modifying the sensitivity of the receptors, and an actual stretch or tension on the muscle which would tend to lead to a deviation from that intention. The consequent reflex response brings about a toning-up in the relevant muscles just sufficient to match the disconnecting or deviating forces and thereby ensures the carrying out of your intention.

In this part of the series we'll begin to bring together the inherent instability and the response into a new picture of how it is possible to maintain a supported uprightness without losing any of the freedom to take that uprightness into various activities. But first, we must explore a little deeper the extent to which misconceptions are woven into the fabric of our common cultural and scientific approaches to the body and being. In the most obvious sense, our uprightness depends upon something being done to deal with the instability, to prevent the de-stabilizing of the uprightness into falling. The something which gets done, however, depends very much on where you are trying to get.

If you are trying to *get up* out of the instability, up from the downward pull of gravity to some balanced posture, you will probably end up using yourself in a manner similar to a jointed mast which is secured from falling by muscular guy-ropes. The object is to get the parts of the mast nicely stacked up, but if the mast tends to fall in one direction, you simply tighten down the guy-ropes a bit on the other side of the fulcrum and bring it back into "alignment". If parts begin to sag out of place you strengthen the "weakened" retaining muscles to pull the parts back into place. Variations on this general theme of alignment and posture are common fare in fitness programs and in books and teaching all the way from grade school biology to Gray's Anatomy.

The flaw in this approach is contained in its underlying point of view. For, in the process of attaining this good posture we only look for the results we are trying to gain and don't usually notice what we are doing in order to achieve these ends. We have, in fact, been so conditioned to operate objectively — to understand and manipulate the mechanisms of things outside us — that we also operate ourselves as a series of objects whereupon we use one part of ourselves to do something to another part. Because of this narrow focus of attention on the parts we're doing something to, we don't consider the effects on the parts which we are using to do the something with, let alone appreciate what goes on with our being as a whole. Out of sight, out of mind, maybe, but certainly part of my use; and unfortunately, what I don't know definitely can hurt me.

Take the example of the abdominal muscles "supporting" the abdominal organs (see also Part 2 of this series). If we feel that our abdominal organs droop and sag out under the influence of gravity we will naturally feel that something must be done to support them and we will tend to use the abdominal muscles for the job of pulling them up and holding them in. In effect, we reach down with these muscles from their attachments to the ribs, get hold of the organs and lift them up. We are, in other words, hanging our organs like a set of weights from our ribs. If we lift up on the organs with a particular amount of effort, so we will be pulling down on the ribs with the same effort. Not only are we pulling downwards, but also narrowing inwards since two of the abdominal muscles run diagonally from the pelvis to the ribs. The same ribs, remember, which seem to have something to do with breathing; breathing which requires that the ribs expand upwards and outwards — exactly the opposite direction to where we are pulling them. Thus, it is not possible to use the abdominal muscles in a holding-in sort of way without interfering with your breathing.

No one, of course, is born with organs too big for the container of their torso. There are only two reasons why the belly sticks out. One is that too much is going into your mouth and no amount of holding in of your stomach is going to correct that. The other is that we are squeezing the space in the container of the torso and forcing the organs to escape that pressure out through the largest open space there is — the abdomen. Ironically, it is the abdominal muscles themselves which create much of this pressure, squeezing in the waist and pulling down and narrowing the chest so that the organs want to push out the belly. The more we hold it in, the more the organs are pressured to come out, so that the more we are forced to hold it in. The moment we let go of holding the organs, they come spilling out and we have reinforced our belief that we need to hold them in.

We also reinforce our belief in the concept — no, more than that — in the experience that our muscles are too weak to do the job and in the desirability of strengthening them. The more we hold in, the more pressure is on the organs, the more they want to come out, and the more we feel that we are barely containing them. If only we could get those muscles toned up and strong so that the job would be more effortless, like a strong man who can easily carry a load instead of a weakling who struggles along with it. With a bit of reflection you can appreciate the twist in this thinking. You can never be strong enough. You can train your muscles to be very good at holding your guts in, but the stronger they get, the more they pull down on your ribs; the more pressure they put on your organs, and the stronger they have to keep getting. Such a waste of energy and it certainly isn't doing your organs any good to be squeezed, nor your breathing any good to be held.

To get back to the flaw in the point of view, as long as you are concerned with your organs coming out you will tend to do something to hold them up and in. If you notice the tension of the holding up and want to free it, you'll have to let go the holding up, that is, let go your organs from above. When you release a holding up the only place things can go is down. We have given them over, relinquished them, abandoned them to gravity. But we don't want that so we pull it all back in again and find ourselves stuck between being "free" but deplorably down (slouched, collapsed), or being up but hopelessly held (tense, stiff, tight).

36

With this in mind, you can see how the flaw in the approach is in the point of view of what we're looking for. If we look down the muscles to the result we want in our abdomen, the only possible freedom is a freedom down. It is built in to the whole approach; it is the way we are directing our attention. If, however we can turn around our point of view to look up the other way at how we are pulling down on our ribs and our breathing, we can release that pulling down thereby liberating our breathing and opening up the container so that there is less pressure on our organs and hence less need for "support", whereupon we can release even more and get even more breathing, more freedom, more lightness, and more openness.

I'm going on at some length with all of this because I think it's important to come to a deep understanding of common patterns of use and to be able to hook up the physical manifestation of the habits with the underlying beliefs and operating strategies. Chronic slouchers, for instance, have had their memory cells permanently seared with admonitions to "stand up straight," "get your chest up," "pull your shoulders back!". They have responded by habitually holding up their habitual slouch. The slouch isn't gone just because they have straightened up. It's merely been covered up so that it appears to have gone. The moment the holding up is let go the slouch reappears. But, in one sense, the sloucher is not really holding up anything, that is, the actual doing is not a holding up of the front. It is a pulling down with muscles in the back in order to raise the front which has been let drop.

This leads us back to the image of a jointed mast with muscular guy-ropes. A jointed mast is inherently unstable. Even if perfectly stacked up with each part exactly over the other it will not stay that way. The mast will tend to buckle and bow. This is especially true if we have a muscular guy-rope attached at both ends to the mast over a joint and we pull up on the lower end (as we might to hold in our guts) which in turn pulls down on the upper end. If this tightening is not to bow the top of the mast over to one side, we must balance it with a tightening down on the other side, so that we now have two pulls-down on the length of the mast. Unfortunately, the more pulling down, the more the mast will tend to buckle out from under the pressure and the more need there will be for tightening the guy-ropes and the more pulling down there will be. It's a vicious circle. From this vantage point it must be acknowledged that it is not gravity which pulls us down, but our own muscles. Gravity is not the enemy; if we look the enemy in the whites of the eyes, it is us!

Think of the skeleton, jointed together and completely free, nothing to stop flexibility or free movement. What restrains and restricts us from our full free ranges and motions is what we do with our own muscles. For whatever reasons, we're holding on to our own skeletons for dear life and can't let go. The problem is not that we are too weak and thus need strengthening exercises. It's that we're far too strong at using our own muscles to grab hold of ourselves to keep from falling over and to pull ourselves into shape. "C'mon, get a grip on yourself, shape up, pull yourself together".

We must learn the skill of how to let go with muscles, to release ourselves. Not in any old direction, especially the beguiling direction of down, but the skill of releasing ourselves in the direction we wish to go. First and foremost that direction of release must be up into the freedom of our length. We need to find not where are the right positions to go to, but what we are doing to stop ourselves from being freely where we are. In particular, this means to find how we pull down and weight ourselves and then simply release up out of it.

37

How, you might ask, is it possible to conceive of releasing up? And up to where? Does this not defy gravity? If it doesn't seem to work to try to reduce the inherent instability in order to achieve a stable balance, could the instability be there for a reason? Well, I've taken a rather roundabout path, but we can now begin to see how the very nature of muscle, with its sensitivity and responsiveness, meets our inherently unstable nature to create a higher level of stability.

To get the whole picture we must step back and take in the whole creature at once. I have to use words to describe this process to you. Words work in a linear fashion, one after the other to create ideas and then progressions of ideas. So, I have to express this to you as a sequence of events as if each one causes the next. But be aware that it is not a situation where one thing happens, then this, then another, but rather it all happens at once, in a gestalt, the moment we release up out of the interfering way of operating. Thus, as I describe each aspect to you, keep it in your mind while you add the next and the next until they all come together as a whole pattern.

Let's begin from the state of being just before the moment of release. We'll use the example of the person above — someone who gets up out of his slouch, raising the forward falling of his chest by means of tightening down the muscles in his back. It doesn't matter whether this brings the chest up too little or even over-raises it into an arch backwards, for regardless of the degree of alignment attained, the goal of balance (reducing the instability) will have been achieved by the simple expedient of having muscularly stiffened his torso so that there is no flexibility nor any moving parts to be unstable. A solid block cannot slouch. Of course, in addition to the loss of flexibility he will have compressed his organs, restricted his breathing, weighted and heavied himself, etc.

If this person could become aware of where and how he has fixed and pulled himself down and could manage to change his operating direction and release up out of these pulls, then a number of things would happen all at once. He will have let go the stiffening and so his various bones will be able to move on each other — he will be more free and flexible. With this freedom will come a greater range of use and thus a greater repertoire of potential movements. He will have released up the pulling down and so will feel lighter. When someone has been heavying himself, he gets used to using more effort to move the heaviness around, so if he feels lighter he would also be able to direct for less effort in activity.

As he releases the narrowings and pulls-down his spine will lengthen and his back and chest will expand giving him more openness. With this openness comes more room and a facilitating of all the organ functions (including circulation, digestion and breathing) all of which will give him a more vigorous metabolism and the vitality to use all his new openness and flexibility. Furthermore, as a bonus from the openness will come more sensation and awareness, more being in touch with himself, and the ability to be more present in whatever he is doing.

All this is wonderful and most people could use a lot more of the above qualities than they normally give themselves. But these are only agreeable side-effects to the primary occurrence — the one which must be present if all these others are to be maintained.

At the moment of release up out of the pulling down, the parts of the torso become no longer bound into one chunk and the inherent forward instability immediately reappears as a tendency to stretch the ungripped muscles up the length of the back. This satisfies one of the two conditions for the stretch reflex — the stretch itself. However, the re-appearance of this instability and the release of the tightness does not lead to slouching as it had before because the other condition — the intention — has changed.

No longer is the intent to release a holding up thereby giving permission to a letting go downwards whose stretches will be unresisted by the stretch response since they match the implicit intent. Now the intent is to release in a different direction — upwards. This new direction sets up the sensitivity of the reflex so that all the lengthening and opening in the world can take place in an up direction in accord with the intention, but any further stretches from the instability which would manifest as a falling forward will be met with the activation of the muscles to resist any further stretch.

So far, so good; but we must go farther. What we have here is a mechanism for doing work much like a waterfall where water flows along a river and falls over a cliff. The potential energy of the water high up on the plateau can be made to do work when it falls: to grind corn, to saw wood, to generate electricity. We have a similar, though less obvious, situation in ourselves. We are a culmination of the sweeping tendency of evolution for creatures to come up onto the land and progressively more and more up off the ground to the point where humans have come as much upright over the ground as it is possible to conceive. The only difference in this analogy between the waterfall and us is that the river has a constant stream of water which it can keep dropping off the cliff. We don't have a steady stream of heads we can drop off our necks so we need a means to keep restoring the uprightness so we can keep drawing upon its potential energy.

We can see how it works if we separate the inseparable for a moment to look at the components of the stretch and the components of the response. For instance, in the muscles running up the back of the neck to the head the stretch has two components. There is an upward pull on the muscle from the forward instability of the head since the back of the head would go up if the head nodded forward. At the same time, and exactly equal, there is a downward pull on the length of the muscles from the mass of the parts to which the muscles are attached at their lower ends as these parts fall toward the centre of the earth. These are not two separate stretches but two ends of the same stretch so, of course, they're equal. You can't pull on an elastic band and only feel the pull back on one end.

When all the lengthening and opening in an up direction has occurred at the moment of release, the muscle will respond with a resistance to any further stretching which would lead to falling over. The amount of this response will be exactly equal to the amount of the stretching and will automatically vary with the stretching should it change. This response, like the tension on the muscle, also has two components (bearing in mind that these are two ends of the same response and ultimately inseparable).

The first component is the toning-up which pulls down the length of the muscle just matching the stretch upwards from the instability of the head forwards. This is the factor which keeps the head up there so we can keep using it. Since the muscular response is exactly equal to the instability it does not let the instability increase and manifest as falling (the muscle does not relax and get longer), nor does it eliminate the instability by reducing it to a "balance" where the head sits directly over the joints of the top of the neck (the muscle does not shorten and pull the head back). Thus the instability is still there the next moment still exerting stretches on the muscles and still activating the reflex response.

This is the component which ensures that the process carries on and that we can continue drawing on the potential energy of the height of the head.

The real payoff, the reason for the evolutionary infatuation with upness, is the other component. The toning-up of the muscle in an upward direction literally draws upwards on the bones to which that muscle is attached below. This drawing upwards is of course exactly equal to the mass of these parts which are stretching the muscle in a downward direction. In other words, the muscles suspend the parts underneath them. The net effect is that the body is hung in a web of musculature from the head. It is not then strictly a support system, but rather a *suspension system.*

Here, the skeleton is not a set of building blocks for us to lumpenly sit on, but a spacer system giving us length, breadth and volume and spreading the muscles apart into stretch whereupon, if ungripped and open, they become exquisitely sensitive to our inherent instability and to changes in equilibrium and they respond to this sensitivity (and our intention to remain released and open) by suspending the skeleton down to lightly rest on the ground upon which it is supported and from which it extends upwards as a long spacer to spread apart the muscles which in turn respond to the stretches by suspending the body lightly down to the ground upon which it is supported...

Thus we are gravity processors — if we can get out of our own way and allow the system to operate as it is designed to operate, and as it had been operating for millions of years before we came to feel responsible for doing our own posture. Gravity doesn't pull you down, you pull yourself down; it is gravity itself which keeps you up, if you don't try to help it.

You can understand now why any attitude which regards our inherent instability as a problem to be solved and which solves it by attempting to reduce the instability to a stable balance point dooms us to all the misuse we explored in the first half of this article. As soon as you pull your head back to balance on your neck and straighten up by pulling down on and holding your back, you kill the whole reflex activation since there is no longer any instability to activate it. The head is in balance and the back is one solid chunk. If you don't have the skill to allow the subtle in-built reflex to be activated and work for you, you have no option but to take on the job of upright posture yourself. But, of course, you have already done that by pre-empting the reflex when you went ahead and postured yourself into straightness and balance.

The beauty of a suspension system is that it maintains uprightness without in any way fixing joints or restricting freedom since it doesn't have to grab bones to get anywhere or hold them to stay there. Quite the opposite, we get this suspension by releasing ourselves. Because of the very nature of muscle tissue and because of our inherent instability and because of the existence of the stretch reflex all the rest takes care of itself.

This whole concept is quite a mouthful with vast repercussions and implications. It is not easy to grasp (sic). It has only been about four years since I discovered this conceptual framework after years of trying to fit the experience of lightness and openness into the old mechanisms of kinesiology and physiology. It has taken me a long time to become as clear about it as I am now. Through every ongoing change in myself and in the insights of teaching, I come to understand it more deeply in experience and hence in conception.

We'll go on in future parts of the series to explain this more (especially in how it is all organized by the primary freedom of the head to lead) and we'll see it in operation in our various areas and functions. For now, just allow yourself to absorb it. Ponder it, but don't try too hard to figure it out. It is innately paradoxical, especially in relation to our normal way of understanding and you would do well to not try to break apart the paradox and reduce it to some objective mechanism appreciable by your regular way of thinking about yourself. This is a truly subjective process, that is, the only full understanding is in the experience itself, and the experience is undeniable. You cannot use the old tools of analysis on a process which is ultimately not separable into parts. Remember the old joke about the Maine farmer who, when asked by a tourist for directions, replied, "*You can't get there from here*".

In Our Own Image

Sixth in a Series on Human Design and Function
Reprinted from *The Alexander Review, Vol. 2 No. 3, Sept. 1987,*
Subsequent editing, June 1993

More Suspense

In the last part of the series I contrasted the more conventional point of view of a support system where we are stacked from the ground up like a mast or a set of building blocks, with my new model of a suspension system where the skeleton and all our parts are hung in a web of elastic suspension from the head to dangle down to the ground whereupon we lightly rest and from which the skeleton functions as a spacer system extending upward, freely spacing and spreading out the muscular and connective tissue into a springy and elastic web which suspends the skeleton down to the ground from where it extends up… and so on. Now, in this part of the series, I will carry on to explain more about the suspension system and its implications.

First, to make it easier to visualize this paradoxical system, allow me to show you a model. The actual model I built is made of little plastic bars hinged together with small bolts and strung with thread. I've made a drawing of it above. If the drawing seems at all ambiguous, you can clarify it using your hands. Make a "V" shape with your thumb and index finger so that they are roughly at a right angle to each other like the plastic "V" shapes, folding up the other three fingers into your palm. Hold the "V" of one hand upright and place the other hand upside down so that the top one is rotated 90 degrees over the other. Holding them like that, imagine tying a thread from the tip of one thumb to the tip of the other thumb, and then onwards from that thumb to the tip of the finger beside it, from there to the next fingertip, and then back to the first thumb again so that you have a line of imaginary thread running in a rectangle around the outside of the

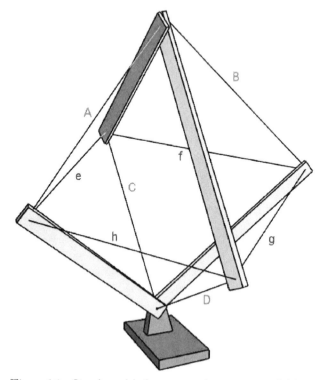

Figure 6.1. Simple model of a supported system suspended in a web of tension

model. Now tie a thread from the tip of each thumb to the other hand where the thumb and finger meet at their base, and do the same from the tip of each finger to the crotch of the thumb and forefinger of the opposite hand.

One of the first things you'll notice about this model is that the upper "V" is not sitting on anything solid; it is suspended in mid-air by only its connecting threads. It is simply the tension (or resistance to being stretched) of the threads which keeps it up. Most people's first impression of this device is: "That can't be; how can that work!". But try to imagine how it would be possible for the top piece to fall. If it tipped over in either direction toward one tip or the other of the lower "V", it would instantly be met with the resistance to stretch of the thread going from its centre down to the tip of the other arm of the lower "V". If it began to swing down to one side such that one of its tips would hit the ground, it will be met with the resistance of the thread which is attached from its other tip to the base of the lower "V". Finally, the four threads going around its perimeter would have to stretch in order for it to fall directly downwards.

This is not a suspension in the normal sense where something is hung from something else, as a demonstration skeleton might be hung from its stand by a hook coming up out of its head. We are obviously not suspended that way. There is no golden thread extending up out of the head suspending us from the sky. Nonetheless, a suspension is the best term for the supporting system of this model — a suspension internal and integral to itself. Indeed the upper "V" is not hung from above, but is hung *by the lower "V" from underneath!* And at the same time the top "V" suspends the bottom "V" since the lower two bars cannot fall outwards from their central hinge because of the tension in the two threads coming down from the centre of the upper "V". At the same time, the bottom "V", lightly resting on the ground, projects up, spreading apart the threads so as to erect the suspension of the top "V". In other words, everything suspends everything else.

The bars, of course, represent the skeletal system and the threads represent the muscle and connective tissue systems. The length and stiffness of the bars act as a spacer system to spread out the threads and keep them under tension. The resistance of the threads to stretching and becoming longer is what keeps the bars from falling. But it is not a static system, for there is an instability due to the freedom of the bars at their joints (in this model, particularly the joint which hinges the two bottom bars). Here is the real beauty of the arrangement — the more the bars fall outwards from this hinge, the more they pull on the threads which keep them from falling. In fact, it is the falling of the bars which expands and erects the whole suspension in the first place, as we saw with our own structure in the last article. If you were to take the two bottom bars and move them inwards on their hinge so that their tips began to come together, all the threads would be slackened and the upper "V" would begin to collapse. As you let the bars go out again into an expansion, they stretch the threads out taut and the upper "V" rises into its suspension.

Many of you will recognize this toy as a variation of Buckminster Fuller's "tensegrity structure". That is, a framework which depends for its structural integrity on tension rather than compression, or to put it more precisely, on both tension (the resistance to stretch of the muscles and connective tissue) and compression (the resistance to deformation of the bones) at the same time. Such a structure is an inseparable whole, every part contributing to the system. Each connective thread shares part of the load of the whole and every thread is essential — cut any one and the system will collapse in one direction or another.

Likewise, the freedom of every joint is essential from which comes the instabilities and consequent stretches which erect the suspension.

In order for this model to move or change shape in any way, some of the threads (if they were real muscles) would have to let themselves get longer, some would have to get shorter, but every one would be involved in some way or another. There are no "prime movers" in a system like this. It is not a system where one muscle does the work by contracting and another graciously relaxes to allow the movement. It is meaningless to speak of "agonist" and "antagonist" muscles in such a total system as this. Those muscles which release to become longer are just as much involved as those which shorten. In some ways more involved, since for most of us it is a much more difficult thing to release into activity than to grab and pull into activity. It isn't even necessarily the case that the muscles which are shortening will be the most active. It may well be the ones which are lengthening that are doing the most work or bearing the largest load. This will depend on the orientation of the structure to the planet, the activities in which it is engaged, and the various loads upon it. More about this in future parts…

There are some differences between this device and ourselves. One is that the threads are not really like muscles. They can resist deformation or collapse of the structure by refusing to stretch, but they cannot change their length. Thus, this toy has only one shape where it can have this suspension — the one pictured. For this to happen all the threads must be taut and they must all be just the right length. If the bars flex on each other so as to come closer and reduce the internal volume, the structure collapses. If they go apart — well, the strings just won't let them because they can't stretch. Even were they elastic threads, it still wouldn't be the same. Elastic threads can stretch, so you could pull the toy into different shapes and the threads would give where necessary. But when you pull on an elastic, its elastic resistance increases and wants to pull it back where it came from. So, not only would you have to use force to pull against this resistance to move to a new shape, but you'd have to hold on to keep it there. In addition, those threads which become shorter during the movement could not take up the slack beyond a certain point and so would hang loose and not provide their share of suspension.

Muscles, on the other hand, have the special talent of being able to be sensitive to tension (stretching and disconnecting forces) and to respond to that sensitivity by resisting any deviation which does not correspond to your intention. This is the muscular function that provides us with support and suspension. Additionally, muscles also have the talent of being able to allow themselves to become longer or shorter, even while they continue to manifest their sensitivity and carry out their suspensory responsiveness. This is what allows us to move. Because of these two talents of muscle we have the capability (if we have the skill) of being able to erect this suspension system not only in our characteristic human uprightness, but also to sustain it (or at least keep coming to it) throughout any movement and into any other positions it is possible to get into. There are thus no "wrong" positions or movements. There may be a greater skill required to release into good use in different positions or movements, and there may be ones in which it is more difficult to maintain a good use, but none that are inherently "bad".

The other major difference between the model and us lies in the fact that we are not strictly a tensegrity structure. Tensegrity structures exhibit continuous tension and discontinuous compression*. That is, their compression elements (in us the bones) are discontinuous like the drawing in figure 6.1 a few pages back. The upper "V" shape in the toy has no "bony" connection to its foot, being suspended in the purest sense of the word. In this it is a bit like the torso of a cat which is suspended between the shoulder blades from its front limbs with no bony joints whatsoever. We, however, do have a bony connection from our feet right on up to the head. Not because the bones are there to be sat upon like building blocks, but because our main spacer direction is directly up away from the planet into our length. I'll go on in greater detail in future writings to explore the details of our spacer and suspension arrangement.

As a result of this main spacer length from head to foot, it is easy to see the body as a mast-like compression structure. The suspension nature of the structure can only be appreciated with a broader point of view which takes account of the inseparable and dynamic inter-relationship of skeletal, neurological, and musculo-connective tissue all activated at once. This doesn't mean that we can't still manage to impose our unconstructive concepts on the design and try to operate ourselves as if we were a simplistic compression structure.

For instance, refer back to the drawing of the model. Imagine that the volume contained within the bars and threads is like the volume of the container of your torso. Pretend that it is facing so that one of the arms of the lower "V" projects out frontwards and that the area just above and to either side of that arm is like the your abdominal area. Further assume you tended to feel that your "abdomen" sags outward and downward when you "relax" and that you should be doing something to hold it all up and in. You could certainly pull your stomach in by contracting the three thread muscles which run from the tip of that arm back and up to the upper "V" (which in this case represents your imaginary chest).

While this might hold up your gut, it would also pull down on your "chest" (especially via the thread running straight up). The pull down will be with a force equal to the holding up. The only way to prevent the chest from being pulled down in the front would be to tighten down the muscle thread which runs down behind to the back arm of the lower "V". Now there will be two pulls down on the chest. While this "solves" one problem, it creates another. Both arms of the lower "V" are being pulled upwards and inwards which

---

* R. Buckminster Fuller writes in "*Synergetics: Explorations into the Geometry of Think-ing*", Vol 1. 700.011: "The word tensegrity is an invention: it is a contraction of "tensional integrity". Tensegrity describes a structural-relationship principle in which structural shape is guaranteed by the finitely closed, comprehensively continuous, tensional behaviors of the system and not by the discontinuous and exclusively local compressional member behaviors. *Tensegrity provides the ability to yield increasingly without ultimately breaking or coming asunder*". [my italics]

Note that he is not saying it is one or the other, rather that the compressional members and the tensional elements cooperate, but it is the behaviour of the continuous tensional flow that guarantees the integrity of the structural shape. Note also that such a structure with its tensional web has an inherent give to it. Forces are absorbed and distributed throughout the whole structure in a springy way.

Compare with this lovely quote on the nature of the architectural arch, a structure made up of only compressional members, albeit in a special arrangement: "What is the principle of the arch? You can call it, if you like, an affront to gravitation; you will be more correct if you call it an appeal to gravitation. The principle asserts that by combining separate stones of a particular shape in a particular way, we can ensure that their very tendency to fall shall prevent them from falling". G. K. Chesterton, "*The Outline of Sanity*" Dodd, Mead & Company, New York 1927

will cause movement at the joint at their base such that they will come together and reduce the volume in the container for the life-support organs. This can only be averted by using other muscles (not present in this model) to stiffen that joint to prevent any bending. In the meantime, if any shortening at all has taken place in the upper threads as a result of their tightening (and this will mean a narrowing in the spacer system), threads elsewhere will be slackened and will either not be contributing their share of suspension, or will have to shorten also to take up the slack. Any further shortening, in turn, contributes to even more strain on the structure which will respond with even more distortion, or more resistance to distortion, and so on...

The end result is that most, if not all, of the joints are held and stiff, restricting free movement. The muscles holding them are shortened and tight, putting pressure on the joints. There is massive pulling down and heavying, which leads to more holding up. And there is an inevitable loss of internal volume bringing about curbed organic function, not to mention that the whole thing is an awful lot of work. All of which is completely unnecessary, since it was doing quite fine when we allowed its in-built suspension nature to express itself before we interfered by trying to help out.

A prerequisite of such an expansive suspension system is its inherent instability. Whenever there is instability (and we as upright humans certainly have that), any system operating as a compression system stacked from the ground up will naturally need to be concerned with balance and alignment, usually in the sense of verticality and straightness. The more up over itself it can get, the less likely it is that parts will fall off each other. It is thus inevitable that anyone operating in this mode will be looking for some variation of alignment as an important element of their posture. This is one of the reasons why people's sensory awareness of themselves comes to be dominated by positional feelings (straight, bent, leaning, forward, back, in front of, behind, etc.).

When operating in the mode of a suspension system, though, the goal of alignment becomes irrelevant. Anything which is hung is automatically dangling down into an alignment, each part or area just over the next, unless for some reason we are holding parts out of alignment. That is to say, alignment is a side effect of suspension. Or, to put it another way, alignment, as any kind of desirable arrangement, is an unconstructive concept.

This is because any concept of alignment as desirable will tend to define an end to be gained to which we will, consciously or not, attempt to align ourselves. It then becomes the habit, the problem, the interference from which we are trying to escape rather than the solution to a problem. In fact, where our parts will end up, when we release into this expansive suspension system, will depend on the sum of all the pulls and stretches on them such that each part will find an equilibrium of its own with its neighbours. The net result, in any positional sense, will be the sum of the nature of the parts themselves, whatever act we are involved in at the moment, and of how totally and deeply we managed to expansively release ourselves. When we take care of the means to our general and overall use, our alignment will take care of itself.

If we can learn to organize ourselves so as to open to this suspension by freeing up out of our habitual pulls down or collapses, by releasing so that the head leads and the spine lengthens, by allowing a widening and expansion in the torso, and a lengthening along the limbs, then balance too takes on a whole different character. Recall from the last article how stiffening and fixing eliminates the instability of the structure by making it into a solid block (or reduces it to several solid blocks tottering on top of each other if the stiffening is partial).

A solid block either is in balance or it's falling. Thus we are led to think of balance as a place, a particular position (usually the one we are familiar with, since we are not falling over). In fact, often in our striving for balance and alignment (or at least what has come to feel aligned), we end up in some place we think is upright but actually far from it, so that, ironically, our pursuit of balance is exactly what unbalances us. Think of the contortions to which some people will go when asked to "stand up straight". This imbalance is what forces us to hold on, which in turn stiffens us so that we have little flexibility with which we can adjust to changes in stability. It would be funny if it didn't lead to so many problems.

One of the requisites of a well-supported poise is a broad and even contact on the supporting surface, as we have seen in previous parts of the series. When we get involved in the pullings down and holdings up of seeking an ideal posture, we inevitably lose touch with our actual state. In practical terms this means not only selective inattention to aspects of our kinaesthesia, but out and out distortion. It always astounds me the high percentage of people who habitually stand so that they have more contact on the front of their feet (or more weight, or more pressure on the front as they might describe it). Quite a few also stand with more contact on their heels or to one foot or the other, but to the front is definitely the most popular, so I'll use that as my example.

What this uneven contact means is that the person, as a whole, is leaning forward and must hold on in at least their calf muscle area to prevent falling. Unfortunately, that holding pulls up on the back of the foot, reducing the heel contact while pressuring the front of the foot. Now, in one sense, the person is receiving valuable information — "there is more pressure on the front = I am leaning forward = balance is back onto an even contact". However, the sensation is a product of his habitual use and as such he may have noticed it but doesn't realize the significance of it and so cannot be responsive to what it is saying. It is this sort of strange bind which forms the basis of the unreliable sensory appreciation we hear so much about. (It is, of course, the appreciation which is unreliable, not the senses — see Chapter 5 of this book, "*Teaching Reliable Sensory Appreciation*")

If such a person could free himself into a better use and allow the suspension system to operate, he would immediately find himself located in space. Even if he does this from his habitual lean forward, at the moment of release, as he frees up into his length, a concomitant release of the suspension down to the ground would occur and he would let go onto an even support on the ground. He would feel the ground as a tangible reality. He would feel grounded. He would also begin to fall since he was out of balance but holding himself from falling when he released. But in the moment of release and grounding, he would begin to feel the actual *process* of unbalancing, and would be able to realize his imbalance and correct for it. That is, balance (or lack of) becomes an active *experience*, not an abstract idea. We begin to perceive equilibrium as a dynamic and ongoing process — a continually re-discovered empirical phenomenon — rather than a static place or position. We begin to get back into touch with ourselves.

In Our Own Image

Seventh in a Series on Human Design and Function
Reprinted from *The Alexander Review, Vol. 3 No. 1, Jan. 1988,*
Subsequent editing, June 1993

It's All Over Now

In the last two parts of this series I've been sketching a model — the suspension system — to explain that undeniable Alexander experience which arises when we perceive our interferences and release up out of them into length and width. This system suspends us into a light and supported equilibrium and opens us into free breathing and organic function all without in any way restricting the freedom of the joints or our ability to go into activity. Just as it takes a radical shift of approach to allow the experience, so it takes a point of view shift to sense how such a system works and the benefits it bestows upon anyone who can access it in themselves.

It has been traditional in anatomy to name parts, divine their local function and try to see the bigger picture by taking these individual functions and arranging them in ever larger systems. Some muscles are postural and help the bones to bear our weight and hold us up; some are for movement and move this weight around; others do organic functions, the diaphragm for respiration, the heart for circulation; and these organic functions in turn provide the energy for the muscles to keep us up and move us around; and so on.

In this part of the series I want to look at how we need to make a shift from a point of view which begins with the parts and tries to use them with all their separate functions to fashion a whole to a viewpoint where we begin with the whole and its larger systems and bring in local areas and parts to show how they co-ordinate with each other and thus fit into the whole. An approach like this constantly illuminates and reinforces the wholeness which we are, rather than the collection of different parts and functions which we must integrate.

This is exactly the nature of the suspension system — one inseparable system made up of every bone and every muscle in your body (including the diaphragm) and all your connective tissue, with a not inconsiderable part of the suspension supplied by the organ systems. Not only does it not make sense to speak of antagonistic and agonistic muscles, but it doesn't really make sense to separate the muscular system into single muscles except for the convenience of locating and describing them. From this point of view the whole muscular system is one flow of sensitive, responsive tissue, surrounding the joints and streaming in and among the bones of the skeleton. A tissue, you will remember, which is sensitive to stretch (instability and loading on the system which manifest as disconnecting forces) and responsive to that stretch by resisting it. We haven't the space here to go beyond a certain level of detail, but I will describe some of the major directions of this flow of muscle.

One of these we've explored already. This is the set of muscles — the so-called extensor system — which deals with the major inherent instabilities of our upright body. The torso is unstable in a forward direction, hence the massive flow of responsive tissue, the two columns of *erector spinae* muscles, which run down each side of the back of the spine to the sacrum and whose job it is to resist stretch so as maintain the integrity of the length of the torso. Because of the torso's instability and our tendency to de-stabilize ourselves

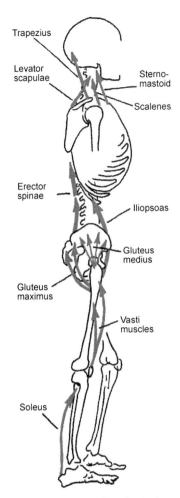

Trapezius

Levator
scapulae

Sterno-
mastoid

Scalenes

Erector
spinae

Iliopsoas

Gluteus
medius

Gluteus
maximus

Vasti
muscles

Soleus

Figure 7.1. (right side view)

when we reach forward, pick things up or bend in a monkey-like way, instability also manifests most commonly in a forward direction at the hip joints, backwards at the knees (which is the only direction in which connective tissue ligaments do not limit the degree of knee movement) and forwards at the ankle joints (which is also the direction of locomotion).

Hence the flow of tissue continues downwards from the spine to spread apart at the sacrum, across the back of the pelvis and hip joints to the back of the femur as the *gluteus maximus* muscles. If we follow this direction of flow through the bone, it re-emerges from the front and the outside of the thigh as the *vasti* muscles of the *quadriceps* which run down into the kneecap and onwards to the top of the lower leg. Passing through the bones to the back of the lower leg, the flow continues downwards as the *soleus* muscles across the back of the ankle to the heel.

The hip joints are ball and socket joints and when we are reasonably poised on our legs there are also other directions with their own flows of responsive tissue branching off from the stream of muscles down the back. In the lower back the two columns deviate forwards to pass alongside the vertebrae, diverging along the inside of the pelvis, over the front of the hip joints and onto the inside front of the thigh, the *ilio-psoas*, responding to instability of the torso backwards.

We are also quite unstable at the hips in a side to side direction, especially when we go into motion. A tributary of tissue flows from the pelvis on either side outwards to the trochanters of the femur (*gluteus medius and minimus*) and the outside of the thigh (*tensor fasciae latae*).

Another branch flows from the bottom and front of the pelvis between the legs out

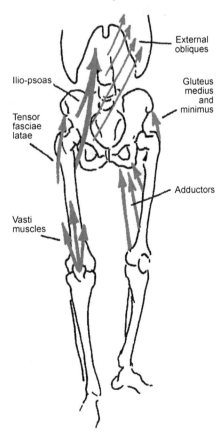

Ilio-psoas

Tensor
fasciae
latae

Vasti
muscles

External
obliques

Gluteus
medius
and
minimus

Adductors

Figure 7.2. Legs (front view)

49

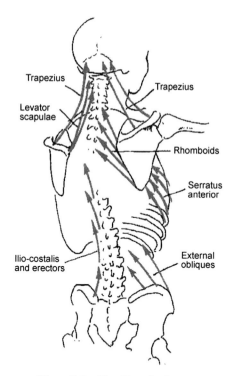

Trapezius

Trapezius

Levator
scapulae

Rhomboids

Serratus
anterior

Ilio-costalis
and erectors

External
obliques

Figure 7.3. Shoulders (back view)

to the inside of the legs (*adductor longus, adductor magnus, etc.*). Instability off to one side will put a stretch along the tissue on the inside of that leg and the outside of the other leg when both feet are on the ground, or if one foot is on the ground the stretch will be along the outside of that leg.

I have not attempted to include all the muscles likely activated by these instabilities, just the obvious and major ones which take care of the erection and suspension of our main spacer length. But this is by no means all of the suspension system, for from this length are suspended all the other parts which thereupon help erect the length.

The upper ribs are hung from the vertebrae by moveable joints from whence the ribs project backwards, around to the side and then to the front to meet together in the middle at the sternum. For the ribs the major direction of instability is downwards in the front and this is resisted by a muscle suspension from the skull (*sternomastoid*) to the sternum and from the upper neck (*scalene muscles*) fanning out onto the ribs. Further back there is a suspension from the lower neck (*iliocostalis cervicis* and *posterior serratus superior*) and the thoracic spine (*levator costarum*). Each rib is then suspended from the one above by a continuation of these flows winding down in one direction (*the internal intercostals*) or the other (*external intercostals*).

The shoulders surround the container of the chest like a yoke on an ox. They have only one bony connection with the rest of the skeleton where the clavicle joins loosely with the sternum. Because of this joint their direction of instability is not down to the sides or down to the back, but rather down and around to the front — the direction of slouching and rounding the shoulders. In a manner analogous to that which suspends the front of the chest this instability is dealt with by a stream of muscle tissue from the head (*trapezius*) and neck (*levator scapulae* — which by the way comes from the same four upper vertebrae as the *scalenes*). This flow continues from the lower neck and upper back as the *rhomboids* which fan outwards and downwards to the inner border of the shoulder blades.

Now things get even more interesting. This flow of the rhomboids from the spine down to the shoulder girdle carries on underneath the scapulae fanning out forwards to the outside of the ribs (*serratus anterior*) thereby suspending them from the shoulders and the spine in much the same way that the upper ribs above are suspended by the *scalenes*. In a similar direction, though higher and more forwards, the ribs are suspended from the shoulder blade (*pectoralis minor*) and the upper arm (*lower pectoralis major*) at the same time that the arm itself is partly hung in a backwards downwards direction (*upper pectoralis major*) from

50

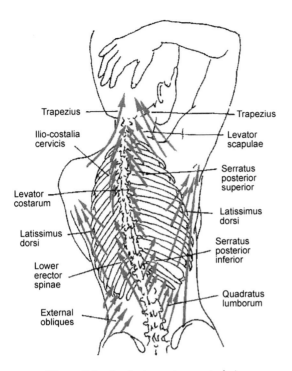

Figure 7.4. (back view, minus scapulae)

the clavicle and upper chest which also are suspended in the same direction from the neck and head through the *scalenes* and *sternomastoid*. This flow of suspension continues from the arm downwards and backwards as the *latissimus dorsi* to suspend the lower ribs, spine and pelvis behind.

The lower spine and pelvis are also suspended from the ribs in the same direction by a deeper flow (*internal intercostals, posterior serratus inferior, quadratus lumborum* and the abdominal muscle, the *internal oblique*, which flows in the same direction as the *internal intercostals*).

In fact the main abdominal wall muscles, the *internal and external oblique*, not only run in the same directions and same planes as the *intercostals*, they are the same muscles. Evolutionarily, the only difference is that there are no longer ribs between the abdominal muscles so they have acquired a different name. Thus, around its sides and front, the pelvis is also suspended from the ribs (*internal and external obliques* carrying on from the *intercostals*) and it is suspended directly in the front from the bottom of the sternum by the *rectus abdominus* which carries on a flow suspending the front of the throat and the top of the sternum from the skull and jaw (*supra-hyoid* and *infra-hyoid* muscles). To come full circle, the pelvis

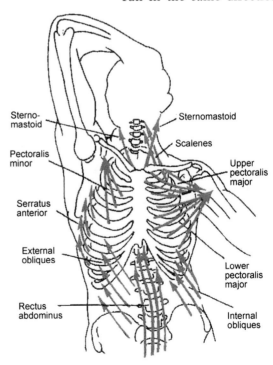

Figure 7.5. Chest and Abdomen (front view)

51

and spine is also suspended from the ribs by the *erector spinae* muscles themselves which come downwards from attachments out on the ribs as well as from the vertebrae.

Well now, let's see. We've got the spine and pelvis hung from the ribs which are hung above from the head and spine. The arms and shoulders are hung from the head and neck and themselves hang the spine and pelvis and the chest which hangs the spine and pelvis below which hang the legs. Everything is hung from everything else just as in the tensegrity model.

The more that we are able to release the chest into width up out of our pulls down and narrowings, the more that it opens up and expands thereby augmenting the erection of the length of the spine which allows us to widen the breathing… You can see how any holding of the shoulders will be a pulling in and tightening around the chest and a pulling down of the head and neck. The more we can release our arms and shoulders, the more they release outwards (like the outwardly releasing bars of the tensegrity device), the less they interfere with free breathing, the more they stretch and further activate the muscular suspensions up to the head and neck and down to the ribs and spine which facilitates the expansion of breathing and the integrity of the length of the spine…

Many of these directions of muscular flow travel on out into the legs and arms, but we'll leave these for another day. However, the suspension system doesn't just stop with the musculoskeletal system. The major organ systems also contribute to it. The entire respiratory system hangs from the head down inside the chest with little in the way of any other supportive skeletal attachment. The suspension of breathing begins directly from the base of the skull behind the nose (the *pharyngeal, palatine* and *hyoid* muscles) and flows down into the throat where it is joined by tissue coming down from the jaw (*supra-hyoid* muscles and the *tongue*). The jaw, of course, and the tongue (which is to

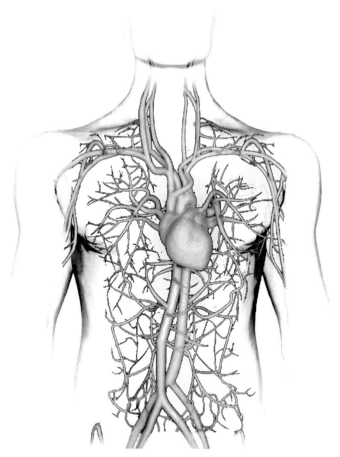

Figure 7.7. The heart suspended in its network of arteries and veins

say the upper part of the digestive system which shares the throat with respiration) are also suspended from the skull.

From the larynx and epiglottis the two passages separate, the trachea continuing down into the bronchial tubes and lungs, the oesophagus continuing on to the stomach. The lungs are not attached to the ribs. When we breathe, nothing gets hold of the lungs and pulls them open sucking in air. We live at the bottom of an ocean of air under considerable pressure which is all too happy to force itself into your lungs pushing them out against the chest wall whether you allow your chest to expand or not. To say that the respiratory system is suspended from the head is to say that it is suspended from the length of the spine. For it is the spine which erects the head thereby creating the length from which the breathing can be suspended. If we pull down or collapse that length, we drop our respiratory length down inside the chest where there is no extra empty space to accommodate it.

The digestive system is also suspended from the head directly and indirectly via the jaw with no other appreciable supportive connections except from the respiratory system as it passes into the chest, through the diaphragm and into the stomach. Out the other side of the stomach the intestines are gathered together by folds of tissue, the *mesentery* and ultimately suspended from the lower back and pelvis. Here again the openness of the length of the digestive system depends upon our ability to erect the length of the spine.

When I say that the respiratory and digestive systems depend from the head with no other substantial supports I meant in a musculoskeletal sense. There is one other major suspensory element — the circulatory system. The heart is a large and powerful muscle suspended in a web of its own arteries and veins. For most of us, our direct experience of the heart is the occasional thumping away in the chest during excitement or exercise. Likewise, our experience of arteries and veins comes from tiny ones bleeding when we've injured ourselves. It takes a leap of the imagination to picture your heart forcefully driving quarts of blood out into every extreme of your body through immense muscular arteries thicker than your thumb under pressure great enough to overcome all your tensing and tightening.

The heart is the centre of a vast and gnarled rooting system of thick resilient tubes all pumped up semi-rigid with blood. This twisting branching root system reaches out past your ribs, through your armpits into your arms. It snakes up into your neck and head, down into your guts, your groin, and your legs, but most of all it deeply penetrates your lungs sending its hungry tendrils into every nook and cranny. Your lungs are almost more blood and vessels than they are air and alveoli. From all these bones, muscles and organs it take its own suspension, defining a place and relationship between all the parts

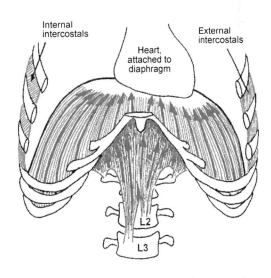

Figure 7.6. Diaphragm (front, ribs cut away)

into which it reaches. It forms a significant part of the suspension of the internal organs, the surrounding container and even the diaphragm which is suspended underneath it. In return, the release and expansion of the chest, the shoulders and our whole length serves to keep suspended and open the network of arterial and venous blood flow.

However, the diaphragm is perhaps the greatest of all the unseen contributors to the suspension system. Originally a back muscle recruited by evolution for the mammals, the diaphragm is so arranged that from all its bony attachments — the lower back, the lower tip of the sternum and the whole lower edge of the rib structure — its fibres head upwards to meet in a tendinous central area to which the heart is bonded. Thus, the diaphragm draws upwards upon all these attachments functioning as a major suspender of the ribs and spine not only when we are breathing in, but also while it's active during expiration.

Whatever particulars of the suspension system I may have left out, it is clear that it ensures every area is suspended and suspends other areas which in turn supplement the suspension of that which they are suspended from. The net effect is support and openness with each function free from interference with and able to augment the functions around it. It isn't even a case of this happens which causes this which leads to that. It all happens at once, integrated and inseparable. So that the release into length *is* the release into width which *is* the release into expansive breathing which *is* the openness for organ function which *is* the intensifying of suspension which *is* the lengthening…

In Our Own Image

Eighth in a Series on Human Design and Function
Reprinted from *The Alexander Review, Vol. 4, 1989,*
Subsequent editing, June 1993

The Primary Control

A new introduction to Part 8 for this edition

The first seven parts of this series were written in one year between January 1986 and January 1987. It was two years later that this final part was published. The preamble to the original published article below explains why there was such a delay.

It is now 23 years since I wrote this article and I would like to say a few words about it from my current point of view.

Throughout the article, I write as if the particular understanding of the primary control, presented there is that of the Alexander Technique. In fact, from my many previous years of teaching in Alexander Technique training courses and workshops for teachers, I could see that this territory of our central coordinating mechanism was one of the most misunderstood of all the fundamental principles of that work.

So my goal with the article was to marshal the facts of human coordination and organisation in order to shed light on both the many inaccuracies I knew abounded as well as to show that updating our understanding of this central part of our nature had big implications for the profession and our way of teaching. I was in a position to do this because I had behind me a lot of knowledge about physiology from my anatomy studies and human dissections and had written *The Body Moveable,* a 600-page text on the human musculoskeletal system ten years earlier. This depth of factual knowledge about how we function combined with years of teaching experience was a rare commodity in the Alexander world at that time.

This article appears to have had some influence on the ideas of Alexander teachers since it was written, and indeed is part of the teaching and reading on a few Alexander training courses and university courses taught by Alexander teachers. But they are relatively few, and even 20 years later I can see that the material in this article is still revolutionary and its implications to the understanding and teaching of the Alexander Technique largely unrealized.

When I wrote it, I knew the material was a challenge to more than one brand of Alexander orthodoxy and was expecting some degree of controversy. I even invited feedback and discussion at the start of the article. In the end I received relatively little feedback — as to why the impact of the ideas in this article was so, shall we say, muted, I'm not sure. Maybe it was because not many people read the article, or they didn't grasp the full implications of it. Or maybe they disagreed, but didn't see why it is important for any work to thrash out what is true and what isn't true in its basic premises or else risk the inevitable profession-wide effects of fundamental contradictions. Maybe it was a little of all these reasons.

In any case, the insights I had at the time which are presented in this article and another written around the same time, *The Rounder We Go, The Stucker We Get* (which comes later in this book), have revolutionized my work and eventually led me in a direction where it made more sense to start a whole new work, *LearningMethods*.

With this history in mind, read on…

Preamble to original 1997 print edition

First allow me to apologize for the interruption between Part 7 of this series and the present article. Both my readers wrote in to enquire what had happened — one to ask that if indeed I was dead could she write the next article instead; the other to say that he was following the series with great interest and would I please get on with it so he could finish putting himself back together. The reason for the hiatus has been a case of acute busyness. Getting an Alexander Technique teacher training program up and running, being on the executive council of STAT (one of the Alexander Technique professional organizations), not to mention a bit of personal life has pre-empted what time I used to have to sit down and write.

Therefore, for this article, with my second reader in mind, I'm departing somewhat from the path I had been pursuing with the others. The territory I am going to explore here has come over time to seem obvious and self-evident to me, but I recognize that it may not seem so to others. I would welcome any correspondence from those who wish to discuss the article (contact me at david@learningmethods.com).

The Primary Control

I've always liked the idea that the principles underlying the Alexander Technique were not just made up by Alexander — that the Technique is not a method he sat down and figured out to cure his vocal problems. Rather, in the pursuit of understanding his problem, Alexander *discovered* the principles (and this is where his genius lay) representing nothing less than the discovery of the way we are built to function. The method which Alexander then evolved from his discovery reflects profoundly these inner principles of our being. Not incidentally, it also relieved his hoarseness and has helped countless others.

Obviously, we are built of a multitude of parts: bones, muscles, organs, nerves, etc. There are many systems of parts making up our structure which provide necessary functions: respiration, metabolism, locomotion, sensory awareness, and so on. Alexander discovered what allows all these parts and functions to be organized into an integrated, coordinated, sensitive and responsive whole — the self. At the same time, he also discovered that which, when interfered with, dis-integrates, mal-coordinates and de-sensitizes us. This organizing factor he called the "*primary control of the working of all the mechanisms of the human organism*"†.

---

† F. M. Alexander, *The Use of the Self*, Page 28, Victor Gollanz Ltd, London 1987. Also note that in the Preface to the New Edition [1941] of the same edition, on page 19, Alexander writes: "…it was my early recognition of the need for preventing what was wrong, that led me to the discovery of the primary control of my use, and *I emphasized this discovery as the all-important one* in my efforts to teach myself." [my italics] and further down on the same page: "…the discovery of the primary control opened up a road by which I could make a safe passage from "idealistic theory to actual practice"…".

I can think of no other "principle" in the Technique which is more central to understanding use and misuse, and no other idea which is more misunderstood (with the possible exception of "*faulty sensory appreciation*" which I write about in Chapter 5 of this book). Even the concepts of "inhibition" and "direction" derive from primary control in the sense that the ways we instinctively and unconsciously direct ourselves interfere with the integrative function of the primary control. When we can inhibit these "misdirections" in favour of conscious re-direction we allow the inherent primary control to restore the wholeness and coordination to our system. Inhibition and direction are the "means whereby" we can cease working against ourselves, but the primary control is the "ourself" we are against — the central coordinating agency built into our very essence and fundament. It is no less than our primordial inheritance as vertebrates.

In fact, a basic innate organization is a principle more primeval than the appearance of the vertebrates, who were its exploiters rather than its originators. The earliest simple creatures with undifferentiated structure developed primitive organ systems, particularly the digestive system, to optimize their ability to metabolize nutrition to usable energy and direct that energy into meaningful activity. The digestive system organized the creature into a tube, opening at the mouth to take in food which is absorbed as it passes along the length of the tube eventually being expelled behind as waste. This simple arrangement, by its very fact, creates an organization of *length* rather than roundness.

Moreover, this length manifests an *endedness* such that it has a mouth end and a tail end. This *ended* length, right from the beginning, had an even deeper organization. It was a length *biased toward the mouth end*. The digestive tube is a one-way system — in the mouth, out the tail — with the mouth end being the important end from the creature's point of view. Its locomotion naturally becomes organized to deliver its mouth to the food. Sensory awareness of the world around it inevitably developed at the mouth end to inform the creature what it was getting into and where the food was. The net result is the genesis of *a head* toward which the whole creature is organized and through which its activity is directed. That is to say, the creature becomes a *directed length*, whereby the head leads and the body follows.

It is important to understand just how literal is this organization of the head leading and body following. It is not a question of the head saying "I want to go over there to chomp on that bone, legs take me there". The body doesn't deliver the head where the creature wants to go by taking directions to walk it or push it. The creature is embodied as an active, seeking length with the centre of its awareness of itself organized towards its head and on out into the world. It then releases that centre of self from where it is to where it wants to be. This letting go of its awareness centre (head) from the rest of itself to where it wants to go is what actually stimulates the reflexes which allow the body to follow in a coordinated fashion. As the head continually disturbs the equilibrium and relationships of the moment by turning, looking, smelling, heading off in this direction or that, the job of the primary control is to initiate and coordinate all the mechanisms which act to continually restore the equilibrium and functional relationship of the rest of the creature to the head.

Imagine an animal at any one moment, let's say a horse in a characteristic stance — in equilibrium with four legs on the ground, head up and turned to look at you like a Stubbs painting. That state of balance and adaptation to the moment in which the horse is standing will change if its head goes down to munch some grass. It will change again if it hears a sound and its head comes up and turns to look around. It will change and need to keep changing if the horse moves off into a run with a toss of its head, stretches its head up to jump over a fence, stops in the next field, then lowers its head to chew some more grass.

That is, the horse's *attention* is normally grabbed through its head since the main outer-directed senses are located in the head, and its *intention* to do something is normally lead by its head whether in sensing, in eating, in movement or in communication. Under most normal circumstances it is the head which disturbs the equilibrium and it is the job of the system then, via the in-built reflexes, to continually re-orient the rest of the body to the head thereby restoring the equilibrium which the head is continually disturbing.‡

The ability of the horse in my example to accomplish even the slightest activity of turning its head to look involves a multitude of necessary responses to adjust the distribution of muscle tone throughout its body, to modify its respiration, to attune its hearing, smell and vision, and numerous other adaptations which flash through its nervous system. If it had to work out how to adjust hundreds of muscles and organic functions every time it turned or took a step, it would find itself up on a pedestal in the park unable to move.

Fortunately for us, over millions of years a myriad of in-built functions have evolved to deal with the zillions of details made necessary by even the most commonly occurring situations of life. Because such reflexes take care of these details, you don't have to. You're free to go about your activities learning to do more and more with less and less interference§

---

‡   As an aside, we need to be wary that from the idea of a constant reorganization brought about by a constant disturbance of equilibrium we don't extrapolate the implication that there is some sort of idealized or abstracted state of equilibrium you can get to that is good - a state which is then disturbed in order to go into activity and can be returned to like a "home-place". Most of us already tend to fix ourselves in postural places. It's very easy when you come to this work to extend that original habit without being aware of it by making ideals of what *should* be happening: "I should be able to stand upright when I'm not actually doing something with both my feet on the ground, weight evenly spread, more or less straight and symmetrical". "I shouldn't put more weight on one foot or cross my legs…"

It is important to be aware not only in our own use but also how in teaching we can easily create this attitude in our pupils by working with them only in upright and symmetrical positions thereby reinforcing the idea that they get all their successes in these "home places". This leads people to cultivate stereotyped positions which they are convinced are good for them. A more subtle version of this end-gaining is to try to bring the home place with us into all our activities: "if I could really use myself well, I'd never get out of equilibrium". This tends to develop into the "Alexandroid" mode of careful correct movement so easily ridiculed by the uninitiated.

These are impositions upon ourselves - us trying to take over the job which the primary control is built into us to handle. We fail to realize (and undoubtedly have failed to fully experience) that our system depends upon the changes and stimuli of dis-equilibrium on an ongoing basis in order to be able to integrate and orient itself. Trust in the innate capacity of our system to right itself is more in order here than us trying to be right.

§   This reminds me of a quote of Frank Pierce Jones" from his notes for the unfinished 15th chapter of *Body Awareness in Action*: "The Alexander Technique is a method for improving motor performance by integrating the voluntary and reflex components of a movement in such a way that the voluntary does not interfere with the reflex and the reflex facilitates the voluntary."

It reminds me also of the quote from a paper read by Dr A. Murdock at the St. Andrews Institute on March 6, 1928 (from the footnote in The Use of the Self on page 50, Gollancz 1987 edition): "Mr. Alexander has built up the theory on which he has based his practice from the

(provided, of course, you don't interfere with this coordination by taking over consciously or unconsciously in such a way as to prevent an integrated and appropriate reflex pattern of response to your activity). Thus the primary control is responsible for ensuring the coordination of the primary functions which form the foundation of your ability to act. These functions include the ability to maintain your support and equilibrium, your readiness to respond, your alertness to yourself and the environment.

Much of this territory is attributed to what are usually called the "postural reflexes". I'm not a great fan of that term, because it tends to imply posture (a fixed pose — or even worse, good posture/bad posture) rather than a dynamic interaction which is the purpose and function of the organization. You can make up your own term, but I prefer to call the whole territory the "poise reflexes". Poise seems to me a much more dynamic word, one is "poised for action" at the point of readiness just before activity. Poise is flexible and changing, not fixed. There isn't good poise or bad poise. Poise just is. It has a quality of centredness, suspended in balance between forces, not being swayed by those forces but able to be in response to them. Applied to individuals it implies presence and alertness but with a quiet serenity.

The poise reflexes provide for (in co-equal order of priority):
  1. supportedness and equilibrium,
  2. ongoing readiness and freedom to move and respond,
  3. openness of organ systems providing energy and emotional response,
  4. spatial orientation, alertness and kinaesthetic sensitivity.

In other words, without the full functioning of these fundamental reflexes, any activity you are involved in will not be able to be well carried out because the foundation is flawed. From this point of view, we're not in balanced posture, then in activity, then back in posture. Rather, we're in an ongoing state of activity whereby these reflexes are being called upon in various ways all the time whether the activity shows obviously visible movement or not. If, because of interference, we don't have these fundamental reflexes working for us we will have to control our system in other ways, generally by holding and bracing in order not to fall down. Thus we shut down various degrees of freedom, readiness, and alertness in order to maintain and protect ourselves with the consequence that we must "do" activity by pulling this protected self into movement with effort — the vicious circle of misuse.

With this brief sketch of a head-leading, body-following organization with its reflexes co-ordinating the whole, let's examine the primary control which somehow makes it all work. Each of us, as we come to this work, acquires some conception of this term, "primary control", both from our teachers and as our own busy little minds worry away at it. It is imperative that the ideas we do assemble are relatively true to the physical and functional reality or we create and reinforce a mismatched myth with the inevitable result of becoming more tangled up in our own misconceptions…

One of the first things we need to be clear about is whether the primary control is embodied anatomically. Does it actually exist or did Alexander mean it to be understood

---

observations of the movements of the body as a whole, and he has made use of lost or unused associated involuntary reflexes with a rare insight, and by recreating them into new conditioned reflexes he has laid the foundation for a new outlook on disease and its diagnosis and treatment."

On the other side of the coin, I remember Walter Carrington saying that one of the pitfalls to be aware of in learning the Alexander Technique is the tendency to get "better and better at being able to do less and less".

as a metaphor of functioning? If it is just a metaphor we can hardly say that it reflects an actual innate organization. And how would those creatures who can't appreciate metaphors manage to be coordinated? If there is an inherent functional organization to our being, there must be some actual neuro-physiological embodiment of the primary control. If it exists, did Alexander discover an actual physical organ or structure hitherto unknown to science? This is, of course, highly unlikely. So we must assume that it comprises some already known elements or arrangement of our being which, prior to Alexander, may not have been appreciated to be central to coordinated functioning.

Alexander doesn't help us much here. He shied away from the details of anatomy and physiology, though did not seem averse to accepting what support scientific discoveries did lend to his work. He talked and wrote a great deal about the primary control from the practical and personal point of view, but not the structural. However, the more I put together my knowledge of human function with my teaching experience, the more I become convinced that an understanding of the basic principle of the primary control is the key to using it well. I say it this way, not because the experiential and practical isn't important — *it is essential to real understanding* — but because the more conceptual side is so deeply misunderstood. *All that follows is meant to be an adjunct to experience. In fact, it has been through experience and taking account of what actually happens that I have gradually been forced to revise my old concepts and evolve new ones.*

In recent years, while travelling about to numerous countries and working with dozens of trainings, I've conducted an informal survey of how various teachers conceive of primary control. The sheer variety of different understandings has astounded me. Conceptions range from the territory of musculoskeletal mechanics to energetic flows, from simple common sense physiology to the anatomically impossible, and not least, to a substantial number of teachers who are unconvinced that the primary control really is primary! But by far the most common interpretation is that it has something to do with the neck, particularly the upper part of the neck — the suboccipital, or even more specifically, the atlanto-occipital area.

Though few admit to it in so many words, there are those who feel that the primary control is the relationship of the head-neck-back in the sense of angles, alignment and position. This is partly understandable as a hangover from old habitual perception, but perhaps we partly acquire this as students when a teacher has hands on our head and says something like, "that's it; no, not that; yes, there, that's it!". The "it" in this case is usually consistently the same sort of symmetrical positional place. Thus people become used to a certain sort of "right" experience from which any departure makes them feel as if they are going down or pulling their heads back because the movement contradicts a reinforced expectation of a certain feeling of "relationship" even though in reality they may be quite free. A relationship of right positions of the parts is patently not what Alexander meant by the head-neck-back relationship.

However, the more common conception seems to be that the primary control is a relationship of freedom, not position; particularly that the neck or some part of the neck must be released in order that "forward and up" and ultimately "lengthening" can occur. This is not difficult to understand when the most common format of "directions" mentions the neck first "to be free", then the head second "to go forward and up", then the back or spine third "to lengthen and widen".

60

What does this mean to "free the neck"? What are we freeing? The holding of the neck? But what does that mean? The holding of the neck joints? Thus, are we freeing joints? Or is it the holding of the neck joints by the neck muscles? Thereby, are we aiming to release muscles? Let us take each in turn and see what we can find…

It is obvious that in most misuse there is a great deal of stiffness in the neck. In addition, there is a uniqueness to the upper neck area in the sense that the joints between the skull and the top two vertebrae, the atlas and the axis, are synovial joints like those in your fingers and elbows. These have a small but completely free range, unlike the disc joints in the rest of the spine which are distortion joints providing a flexibility through the elastic distortion of the fibro-cartilage disks. These free synovial joints provide for a small but open range of freedom for the poise of your head in relation to the rest of you. But, in the larger sense the neck is not a entity to be free in and of itself. Rather, it is the junction between your head and the rest of you. It is a limb for your head; your head's way of getting around. In other words, you don't make neck movements to have the neck go somewhere. The neck moves because your head moves relative to the rest of you.

In light of this, it is fascinating how misuse channels us into certain realms of experience. For example, pulling down creates a sense of heaviness and weight which the person does not identify as his own doing but *projects* as actual weight (the kind measured by scales) which will have to be lifted if he is to get up off that chair. Until, that is, he can release his pulling down and discover he feels much lighter. Did he lose weight suddenly? Did gravity temporarily let up? In a similar manner, a person can *introject* his doing so as to feel it only within himself and thus lose the significance of what he is doing relative to what's around him. Often when a person sits by going back and down to the chair he will feel his legs grab and tighten in the thigh/knee area or the front of the ankle. He will tend to feel that this has something to do with the sitting, but try as he might he'll find it nearly impossible to release this holding, quite unaware that these grabbings are natural and necessary balance reactions to his unbalancing backwards to sit. As he begins to come to a use which releases and opens his whole system he will begin to experience directly the significance the dis-equilibrium of how he uses himself in space.

Equally, misuse tends to give us a joint consciousness. Because we habitually hold ourselves, movement tends to take place in only a few "joint areas" between adjacent braced areas thereby reinforcing a feeling of "jointness". In this way we can easily, over the course of years, become familiar with movement only in these favourite joints areas to the degree that others are entirely lost from consciousness. Further reinforcement of this joint consciousness comes from a diminishing of the proprioceptive "muscle sense" from the muscle spindles brought about by a contracted use of muscles. This contributes to a loss of being present (of feeling ourselves — proprioception literally means the perception of your property) in the muscular areas between joints. Add to this the tendency for discomforting symptoms to appear in localized areas usually where we've been using one part of us against another part of us, for instance the lower back or the knees.

If we're not careful, we as teachers augment this point of view. Often in the beginning stages of lessons when pupils are still able to allow release in only small areas of themselves (still being relatively out of touch with themselves) they will often experience this release as the sudden letting go of some joint area or another. Witness the revelation so many students have when they rediscover their hip joints so much lower than they previously felt. This is "natural" and simply reflects the old habit of perception carried into the new experience. But it is our job to always bring our pupils back from this part or that part, however nice or different it feels, to the larger whole of the self. After all, one of the hallmarks of the experience of an open use of the whole of you is one of having no joints or parts at all. Movement seems to take place as an effortless flow of the whole of you at once in ways which a moment before would have seemed mechanically impossible and in which, even though there is a plenitude of experience, there are no details one can easily pin down.

Take the example of the knee. Because our misuse commonly forces us to brace this joint, the idea of releasing the knees comes up frequently, usually with a direction of releasing them "forward" or "forward and away"¶ But what do we mean when we invite someone to release their knees? What is the knee anyway? Is it the whole knobbly bulge

---

¶ I'm indebted to Paul Collins for the information that to the best of his memory in all his searches through Alexander's writings he couldn't find evidence of Alexander ever using the direction "knees forward and away", though he did talk about "knees forward" in the so-called Bedford Lecture given in 1934. Thus it appears that this direction is a development that originated most likely in one training course or another.

In this context, I've often wondered, why only direct some joints and not all joints? Where do we draw the line and why? You only need to have the experience of whole (jointless, partless) movement once to appreciate that when all goes well you don't need to tell your bits what to do. If, in release, your knees do go forward and away we must realize that it's because they were being held back and in. That is, a release is always a *release from* some state of non-release. As such, the feeling of change is precisely that — a change from a prior state — and not an adequate description of the state now arrived at. To the degree that we identify it as a description of a present moment (especially with any intent to recreate it) is the extent that we lock ourselves into the habit in order to re-experience the same change.

Moreover, we must remember that if we can identify the misuse (holding the knees back and pulling down in the legs) and direct ourselves as a whole up off that pulling down, then to the degree that we actually succeed is the extent that the legs will release and the knee will go where it will go when you're not holding it somewhere else - into the most integrated relationship of the moment with the rest of you. Thus, if we are really saying that the body follows, we don't need to tell it how to follow. If the knees will go forward and away when we let go of our interference, then we don't need to tell them. If they won't naturally go that way, then we certainly don't want to compel them.

Whoever it was in the past "discovered" this direction through their own experience, you can be pretty sure they didn't know beforehand what was going to happen. They probably allowed a response of their whole being and had the experience of their knees releasing forward and away relative to what they were used to feeling. Since they didn't need to ask for it directly the first time, but merely took care of some larger direction and it happened for them, why should they need to ask for it the subsequent times? This is the whole point of the primary control. You take care of the primary lead of the head and open up the rest of you to respond with following in its own way. The details of just how that will occur will take care of themselves. To do anything else is like Henry Ford saying, "You can have any colour car you want so long as it's black", i.e., "Body, you can follow any way you want so long as it's exactly like this!"

Do we really feel our system is so uncoordinated that we need to tell it what to do? And in such simplistic terms too. This only shows how deeply our dis-coordinating habits have affected our sense of ourselves. Even if knees forward and away (to use this point of view) is how the system works naturally in normal symmetrical bending, what makes us think it will work this way if we come to standing while turning the whole body off to the left or right? I'm sure one could make a case for these directions being preserved in complex movements, but the point is to see how it tends to lock us into stereotyped patterns of movement. We have then limited our own range and repertoire by trying to get involved in helping out that which doesn't need helping out.

62

in the middle of your leg? Is it the ends of the two long leg bones? Is it the joint between the bones? Does the joint include the end surfaces of the two leg bones and the ligaments and capsule too? Or is it the junction itself between the bones, which is to say, nothing at all except empty space?

When we say "release the knees", are we not really meaning to release the upper leg from the lower leg so that we are not holding the upper leg fixed to the lower leg or the lower leg to the upper? If so, then we're not freeing the knee joint to let the knee go somewhere so much as releasing one part from another part so that both parts can be free to move from each other. It's interesting to note in this context that the direction "free the neck", is not intended to let the neck go somewhere, but to allow the head to go somewhere — specifically "forward and up" relative to the neck. Or is it relative to the rest of you since the neck is a junction between the head and the rest of you?

Also to the degree that we start to direct joints is the degree that we keep ourselves narrowed in. To release your knees, if "knees" are a feeling in the middle of your legs, is to stay narrowed in on releasing some small area as opposed to broadening out your aware-ness to release one part of the whole leg from another part of the whole leg. To release your neck is to go right in to somewhere behind and below your head (back and down?) in order later on to go somewhere else (forward and up?). To free your head from the rest of you is to already be out into the whole of you allowing a release where you don't have to go somewhere else afterwards because you're there already. We need to realize that, in general, misuse constitutes an already existing pulling in and narrowing of ourselves from which we need to release. To notice that the knees are held and need releasing is to recognize that we've already narrowed in to those parts. But it's not the knees which need releasing. It's that everything above the knees has been pulled down to and held onto everything below the knees. Why not have a direction which recognizes this point of view and is consonant with it in all ways?

Let's go a bit further with this. When you think of releasing your knees, no matter how you are thinking of doing it, is it really the knee joints which you are holding? Or is it even one bone you are releasing from another? There is nothing within the bones or joints which, like glue, sticks them together. But, if we're not releasing the joints, are we releasing muscles? Is it muscles which hold the bones fixed to each other? The various muscles which cross and could hold the knee go all the way up to your pelvis and all the way down to your foot. Does this mean we have to release the whole length of the leg to release the knee? Isn't this similar to the idea of releasing the upper leg from the lower leg, except now we can see that we are releasing bones and muscles all together? Or, to put it more inclusively, we are releasing everything above from everything below; that is to say releasing ourselves up off our legs.

Allow me to go back for a moment to the idea of releasing so the knee can go some-where. The knee, no matter how released in the joint or its muscles, isn't going to go anywhere if you are still holding your hip and ankle since the upper leg can't go anywhere unless released from the torso and the lower leg unless released from the foot. So to think about releasing your knees to let your knees go somewhere doesn't make any sense when you think about it. The hip and ankle in turn can be held by muscles which go right into the foot and all the way up into the lower back. If we are going to release such an extent of muscle to allow the knee to go somewhere, why don't we just open up and release the whole of us, especially since we'd also want all the other "joints" released whereupon we'll have to release all their relevant muscular holdings from head to toe?

This is equally true in the neck. The neck muscles consist of many groups of numerous muscles from the tiny, sensitive suboccipital muscles between the skull and the top two vertebrae to large muscles which come off the base of the skull and extend all the way down to the chest and shoulders. In between, muscles run from vertebra to vertebra within the neck in leaps of varying length. Other muscles from the skull span the entire length of the neck. Still others stretch from the upper neck to the ribs and to the shoulder blades. If to "free the neck" means we are releasing muscles with which we hold and fix, which ones are we releasing? The little ones deep inside? How can that have any effect if we are still pulling down on the head with the large muscles that run to the shoulders and chest? All of the muscles together? Since that includes an expanse of muscles from your head to your shoulders and right down onto the chest, does it really make any sense to think about such an extent of release as a release of the neck?

As you can see, the idea of releasing the neck or the knee starts off making sense from our habitual point of view, but with a bit of thinking it doesn't make any sense at all. Things are further complicated by the fact that if you are releasing your knees one way (forward and away) you'll have to release your hips a different way (back and down perhaps) and probably your spine a different way (forward and up). How do we make sense of this inconsistency or connect it to the idea that there is an overall organization when we have parts going in all sorts of directions? To incorporate the disparity you have no option but to elaborate a concept like "antagonistic pulls" or "oppositions': the knee goes away from the ankle and the hip, the hip goes away from the knee and the head. While this does somewhat cover the idea of lengthening and stretch, it all gets more and more complicated until we're virtually back into the kind of complexity more reminiscent of our habits than a simple and whole organization.

If you change the emphasis from releasing joints and junctions to releasing part of yourself from other parts of yourself, you will realize that to release a stuckness in the "knee" is to let the part above free from the part below. If your hip is fixed then you release your torso from your legs. From this point of view a single consonant direction is possible. You release your head out into the world from the rest of you and allow the rest of you to follow. Your torso opens and frees into length as it releases from your legs to follow your head, whereupon you can release your upper leg from your lower leg to follow the torso and your lower leg from your foot to follow the upper leg. Because your foot is free from being held to your lower leg it is released onto the floor into support or to follow the lower leg into movement. The ground itself tangibly comes up under your feet supporting you in this same direction. In other words, you have everything going in one unified direction, each part releasing up off the one below so that your head really does lead and your body follows. Obviously, it is possible to look at it either way, but you have a choice whether you want to see yourself going in one consistent direction (the same direction which automatically leads up and out into your activities), or a very intricate set of directions for each part needing complex concepts to explain them. And remember, after all, it's the same inherent system no matter how you think about yourself. The only thing that changes is how effective you'll be. Try it out and see for yourself…

64

Hopefully, all the above reasons help to clarify somewhat the primary organization of "head leads, body follows" and how the way we think about it mightily affects the results we get. However one chooses to think about it, the general nature of such an organization necessitates that your system gets sufficient information about how your head is changing its relationship to the rest of you so it knows what kind of following is required. Much of this information comes in via movements of the head as they manifest as changes in joint relationships and muscle lengths and tensions in the neck, but not exclusively in the neck. There are changes and sensitivity elsewhere. You are just more sensitive there because, of course, that's where the first changes are going to occur between your head and the rest of you. In a free system these changes will, however, occur all the way through the rest of you in a graduated way that doesn't suddenly cease at the bottom of the neck. Thus, it not that there is anything important about your neck as an entity, it is the freedom of your head from the rest of you and the information generated from that which your system wants.

This is not the only aspect of the primary control, however. If we take into account the actual characteristic experiences people have when they do open up from their narrowing habits we can begin to discern some other elements of the primary control. Along with the internal experience of more lightness and openness regularly comes a sense of expansion and being quite present in the space around one. Surprisingly often this includes being able to see everything more clearly and sharply as well as the ability to actually perceive one's spatial relationship to groundedness and support and to balance and equilibrium. If we are true to these experiences we shall have to see if we can understand what this implies in terms of *"the working of the mechanisms of the human organism"* and how we need to take account of this in directing ourselves.

Orientation and spatial perception are largely the domain of the so-called "balance mechanisms" — the eyes and the *vestibular apparatus* of the inner ear with its *semicircular canals* and *otolith organs*. All of these, significantly enough, are in your head and as such they do not necessarily give you information about changes in your whole body, but specifically how your head moves. Unlike the information we looked at above from the joints and muscles, these senses are not telling you how your head is moving relative to the rest of you but about changes of your head relative to what it was doing a moment ago and relative to the space around you as well as giving you your sense of space itself.

The semicircular canals give information about how your head is changing its relation to what it was doing a moment ago, particularly in rotatory movements (angular accelerations), including rotation up and down. The otolith organs provide details more about linear changes in head movement and in addition how your head is changing its relationship to the planet. Your eyes furnish data on how your head is changing its relationship to the space around you as well as relative to what it was doing a moment ago. All these combined tell the system primarily about changes occurring for your head — where your head is going, how it is hanging from a moment ago and what it's getting into. With the addition of the information from muscle and joint changes due to head movements relative to the rest of the body (primarily, but not exclusively, as they manifest in the neck joints and muscles) your system can determine whether the rest of your body is moving with the head, staying still, or moving in a different direction.

Let's begin to put all this together. The eyes and inner ear mechanisms pick up information about what your head is getting into and consequently what sort of following your body will need to do. The internal proprioception of the joints, muscles and tendons gives information as to how the head (with its eye and inner ear senses) is changing its relationship to the body and hence where the following is starting from and how it is proceeding. In other words, it's not just what your head does by itself which determines how your body follows, it's also important what your head does in relation to the state of coordination of your whole system at the moment — the actual situation of you and how that situation is changing.

All of this stimulus coming in simultaneously ensures an adequate amount of information for the system to be able to generate an appropriate reflex response of the rest of you to deal with the changes brought about by leading with your head into an intention to do something or to pay attention to something. And it all happens automatically without you having to do it or get involved with the details. When combined with the ongoing feedback from any other sources, notably supporting contact (which is to say what state of support on the ground you are in at the moment and how that is changing), your system is able to continually organize itself to continually catch up to the head which is continually disturbing the equilibrium. If you can freely release your head (with its intrinsic awareness of the world around you) off into what you are doing and allow the openness of the rest of you, you get an endless responsiveness of the rest of you to this endless release of your head out into activity — be it looking, walking, talking, etc.

You can see how each bit of this information is essential in determining and generating the response which will be evoked from your system. If you cut this information off in any way, whether by holding joints, gripping with muscles, or locking your head into the rest of your body, there will be a reduction in the information from various of these sources. You consequently rob yourself of awareness and of the resulting appropriate response forcing your system to adopt a cruder and generally more forcefully controlled response since it doesn't have sufficient information to make a more refined response. The cruder the response, the more restrictive, protective controlling there will need to be. The more of this kind of "control", the more you become blocked off and the more you will have to make crude responses. The more this happens, the more you are forced to make "voluntary movements" since your system is not free to move without you doing something. The more you are forced to 'do' or 'make' these movements, the more you pull around on yourself until you come to feel responsible for 'making movements': "It's fine to *think* about standing up, but it won't happen unless I actually *do* it."

A perfect example of this is in walking. It is all too easy to be in a state of use where you feel that to walk you must shift weight in order to take a step. Since when you are free you feel light and without heaviness, this shows immediately that there is weightedness present and you could profit from looking to how you are pulling down before trying to move. The next interesting question is why do you need to "take" a step? For taking a step in that sense is quite a muscularly done thing.

What or who are you taking it from? And why that leg and not the other? Again, experience shows that when you are relatively open and up off yourself, you don't need to take a step. If you open up freely to where you want to go, you don't even need to choose which leg is going to move first, it will all be worked out for you in a way in which you go off into movement all at once, each part releasing into the activity.

We've now run around enough to return to our original theme and see that the primary control is not a single mechanism but a series of inextricably intertwined mechanisms

66

which together constitute a unified structurally-based and reflex-based factor functioning to co-ordinate and integrate your system to the lead of your head. Part of this factor is the sensitivity and reflex response to movements of the head relative to the rest of the body. This part is primarily based upon the freedom of the neck inasmuch as that constitutes the freedom of the head from the body. The neck muscles and joints appear to be extraordinarily sensitive and from movements of the head relative to the rest of you and relative to space, for instance, arise the tonic neck reflexes helping to co-ordinate the trunk and limbs.

Another component of the primary control is your eyes. Obviously need your eyes open if you are going to do anything in real life. There is also an important reason for having your eyes open while exploring your use. Unless your eyes are open and you are fully present in the world out there, you will not be accessing, nor able to use, the information that comes through the eyes. I'm sure this is part of the reason Alexander made such a big deal of not feeling inwards, thereby fixing the gaze and shutting out the world around. Your eyes are responsible for a significant amount of your ability to balance yourself, a fact which is easy to test. Stand up on one leg and notice that, while it requires consider-able activity of the lower leg and foot, it is not that challenging. Now shut your eyes and appreciate how much more difficult it becomes. Start again standing on one leg, then shut your eyes and as soon as you feel the slightest imbalance, open them again. Does it make a difference? If you open them before you become too imbalanced, you'll find that the moment you open your eyes, you become much steadier.

This is the reason why it is absolutely deadly in terms of use to feel that you can in any way pay attention to your use and get things going by withdrawing your awareness from the world, feeling inwards and trying to focus on what's going on inside yourself. Your eyes may be open but you are not present in space, which is like chopping off one of your legs as far as the primary control goes. It's definitely a tempting thing to do because we all want to feel what's happening and figure it out, but if you remain visually present in the world around you, you'll find it changes the character of your use and leaves you available for your activity as a bonus.

Put this in context with Alexander's admonitions about not fixing the eyes. If you catch people fixing their gaze and ask them what they were paying attention to, you'll almost always find that they were looking inwards and were not present in the outside world. Somewhere he says something to the effect of making sure you don't fix your gaze but look around at things in such a way that you could describe them**.

Think for a moment what that means; how present you have to be with something to be able to describe it. Try it sometime and you'll realize that it is nigh unto impossible to look inwards and try to feel out the release of your neck, a release of your knees of some lengthening of your spine and still be present enough to describe something out there at the same time.

** I'm grateful to Betsy Polatin, a Boston-area Alexander Technique teacher, for helping me out here. I couldn't for the life of me remember where I'd heard or read this one. I looked everywhere because I knew that I'd not made it up. Just when I was beginning to suspect that perhaps it was wishful thinking, Betsy showed me the notes she had made at the previous International Alexander Congress of the master class with Marjory Barlow where Barlow describes how frequently Alexander used to admonish pupils with something similar to this.

There's more to it, though. The inner ear mechanisms also play their part by contributing to balancing, to equilibrium and to the sense of orientation in space (remember the poise reflexes mentioned earlier). Reflexes generated from the semicircular canals, for instance, are directly involved in organizing your body to follow your head as a simple example will show. When you stand up and spin around as quickly as you can 8 or 10 times and then stop, it is the input from the semicircular canals which makes you feel you want to walk off into the direction of the turn. When you spin a child around and set them down, they stagger off in the same direction as the spin. By turning a number of times, the fluid in the canals has regularized to the movement and is no longer moving relative to the skull. When you put the child back down on the ground, the skull stops moving but the fluid keeps on going, a change the system interprets to mean that the head is now turning and so organizes the child's body to walk off following the head. Children's nervous systems are not as "mature" as ours and so they cannot inhibit this reflex response. You and I realize through other input (chiefly the eyes) that we are not actually turning and thus can inhibit the body-following response. Note, however, that it still feels as if you are turning.

To give you an idea of how deeply interconnected are these aspects of the primary control, try this. Have a partner turn quickly as before about 10 times, then stop so that they are looking directly at you. Look into their eyes and note what you see. You should notice that after they stop turning their eyes will track off in the turning direction, then return to look at you and track off again. It is this interlinked reflex organization which allows you to turn your head and have the eyes travel out ahead to see what you are getting into. Try it. Look forward so your eyes are centred in your eye sockets, then suddenly turn your head sideways and note that your eyes will now be around to the side of your eye sockets that you turned toward.

While these three organizing factors, initially at least, may seem to be different mechanisms somewhat divergent in scope and purview, they coalesce into a larger unified function when you consider that they are all concerned with detecting the lead of the head and organizing the following of the rest of the system. But the importance of the head leading is not just buried deep in our reflex patterns. It makes up such an obvious part of our being that we often overlook it. It is this whole biased organization, especially the eyes and inner ear components which give rise to the common perception that you, the conscious person, exists somewhere behind your eyes and in between your ears.

This is not just some artefact of perception. Your head really is the bodily centre of your consciousness. It's no coincidence that when you talk to people you find their head and look into their eyes. You don't talk to their knees or their feet or their elbows. You talk to their heads, that is, *to them*. In fact, when you think about it, the way you know it's a head is because it has eyes and ears and a mouth. You, the self, are up there from that vantage point which is the centre of your consciousness looking out at me looking back at you. Of course, hopefully you're not up there cut off from the rest of you, dwelling in some little point of consciousness behind the TV screens of your eyes in between the headphones of your ears operating yourself like some big machine. Your head is not all of you by any means, but nonetheless it is the centre of you.

There are tremendous implications in expanding the concept of primary control this way. It's a funny thing how the mind works. For years I'd always thought of primary control as "neck based", even while I was giving workshops on human function and talking about these reflex mechanisms of balance, etc. at the same time as, but separate from, the primary control. Eventually, in teaching I couldn't avoid noticing how people insistently went inside to feel out their "directions". As they did they tended to become still and ended up getting stiffer and stiffer even while they were busily engaged in trying to release themselves. But when they really did open themselves up and release they so often seemed to expand in front of me and become present in the world. They opened up to a supported equilibrium and into integrated activity. Slowly it began to dawn on me that there was a connection. It was then that it struck me that if I was thinking of the primary control as the organizing factor "of the working of *all the mechanisms* of the human organism", then the very elements of reflex control of balance and coordination I'd often talked about in workshops were here actually showing up in the moment of experience, but in a quite different way than I'd imagined them. Actually, it wasn't so much that they were different than I'd expected, they were just right there in actuality instead of as abstract ideas.

It had never occurred to me to connect the "balance mechanisms", which were supposedly working their unconscious reflex magic, with any direct personal experience. When I did, I realized that it is not the head which leads, at least not that block of bone, brain and hair. It is the much larger quality of "headness" with which the primary control is concerned.

In this larger sense, headness is you the being. Headness is being aware of the world around you... Because you have eyes and inner ears you are aware and present in space. It doesn't take any effort whatsoever to be present in space. It is simply part of the nature of your being, like your heart beating and your breathing. On the other hand, it does take a great deal of effort to narrow yourself in and cut off from what's around you. We don't notice the effort because we're used to it and besides we're actively engaged in trying to do something. But, of course, that very trying is the effort. If I'm trying to pay attention to freeing my neck, that's the efforting that cuts me off from everything else.

In the broader sense, I don't have to allow my head to go forward and up. I am my head — I go forward and up. The very act of opening up to headness automatically includes "worldness", "spaceness", "orientation" and "up out there" as essential properties of headness unless I exclude them. Opening to headness is also to include openness to the rest of me if I allow it and if I allow the rest of me to respond. I then become the rest of me also, still with the centre of my being in a headness open to the world, but also open to myself. When I am all of me, I am present in myself, a result of which is "groundedness" (and all the freedom that allows), "openness" (to organic function, to the room in breathing, to emotion) and a deeper awareness of the world around me. I don't have to do it — I am it. The fact that I am a body (I am embodied) means I am present and open to what's around. If I am cut off, I am not myself and will inevitably not be fully present and no amount of narrowing to release some part will change that. Only if I can get fully up and "inhabit" my being with all my innate organization and awareness will I be able to fully use my self.

While it's certainly possible to get some organizational response from accessing part of this larger primary control, the neck reflexes alone for instance, it isn't possible to elicit the full coordinated response without the full open excitation and interaction of all the elements. In particular, if you cut yourself off from the outside world by going in to do your directions as releasing one body part from another into some internal length, how can you release the whole of you to what you're doing when there isn't even an outside world to release into? The feedback and responses from the eyes and vestibular apparatus are inhibited, or at the very least, drastically diminished by withdrawing from their appointed domain — the world you live and breathe in. In practice the tendency to concentrate specifically on the neck can manifest as freedom of sorts within your body and a definite going up and opening, but without including the other aspects, it remains virtually impossible to take this open use directly into real-world activity. This restriction of the primary control is one of the chief culprits responsible for that careful, stereotyped manner of "alexandroid" movement so often associated with those studying or teaching the Technique (and for which we are so justly criticized).

Just as Alexander was forced many times to re-evaluate his premises and his approach because of the experiences thrust upon him in his experiments, I figured I couldn't lose by doing the same.†† So I began to experiment with the practical implications of the primary control as a broader mechanism which includes all the above aspects with their associated awarenesses. In practice, I couldn't see any way around the effective use of such a primary control being the ability to get up right past my neck, right up into my head and right up out into the world. And to do this in such a way that as I come up and fully out into the world I allow myself to go right on with the direction, releasing whatever would hold me back, opening out to become the whole of me, free and following my opening to the world around me and whatever I'm engaged in doing...

In playing around, I also realized that this, in effect, is what the traditional directions about "free the neck to allow your head to go forward and up...", etc. were getting at when it worked. It would be possible to pour these insights into the accustomed words, enrich their meaning and carry on teaching them to pupils because now they meant even more than before. However, I also realized that, while they would now mean more for me, they would quite likely still engender in others all the potential misunderstandings detailed above. Just because they meant more for me doesn't mean by repeating them more often or saying them louder my understanding would be accessible to my pupils. It became clear that I would have to find other ways to convey my insights and help others achieve their own. In any case, we all need to do this anyway. But enough for today. Perhaps another time I can go into where all of this has led.

---

†† Consider these quotes of Alexander from *The Use of the Self*, (Victor Gollanz Ltd, London 1987): on page 32, "When I realized this, I was much disturbed and I saw that the whole situation would have to be reconsidered. I went back to the beginning again...", and from page 39, "I set out to put this idea into practice, but was at once brought up short by a series of startling and unexpected experiences.", also page 42 "The fact remained that I failed more often than not, and nothing was more certain than that I must go back and reconsider my premises."

Chapter Three

Teaching
Reliable Sensory Appreciation

Extracts from a Conversation with DAVID GORMAN
Interviewed by Sean Carey in summer 1988

From transcripts, with some subsequent editing

Introduction to this new article

This article is extracted from a series of interview-format conversations that Dr. Sean Carey (another Alexander Technique teacher) did with me back in the spring and summer of 1988 in London, England as I was preparing to open my own Alexander Technique Teacher Training School. Chapter Three of this book, *On Fitness*, is also from the same series of conversations.

In the ten years prior to this I had been teaching all over Europe and North America — individually, in small groups, in residential workshops, in universities and conservatories, and at many different Alexander training courses. This had given me exposure to most of the different training and teaching modalities of the Alexander Technique, so I had many experiences to stimulate me into an ongoing calling-into-question of the work and to spur on the inevitable evolution of my own teaching.

At the time I felt that all these changes were taking me closer and closer to the heart of what the Alexander Technique was all about, and I was looking forward to having the opportunity of applying these ways of working in the training of new teachers.

Of course, in the years after these conversations my work kept on changing as I saw more and more clearly that it was not the physical coordinating system that was the main source of people's chronic problems. The coordinating system works just fine and does not need changing. It was the inner point of view of the pupil — the inaccurate ideas, the misinterpretations of feelings and misconceptions about how the system works — that is really the "seat of their habits". ‡‡

Eventually I saw that this evolution was developing beyond where it made sense to use the Alexander Technique name so it became a new work, LearningMethods. Nevertheless, even in this article ten years earlier can be seen the seeds of the importance of how the pupil sees things and the beginnings of practical pedagogical ways to get at that thinking and help make it more accurate.

---

‡‡ I could not agree more with what F. M. Alexander himself wrote in *Conscious Constructive Control of the Individual* (page 80, Gollancz 1987): "A teaching experience of over twenty-five years in a psycho-physical sphere has given me a very real knowledge of the psycho-physical difficulties which stand in the way of many adults who need re-education and co-ordination, and, as the result of this experience, *I have no hesitation in stating that the pupil's fixed ideas and conceptions are the cause of the major part of his difficulties.* [my italics]

SC: I think it would be useful if we used this session to consider some models of teaching. Dr. Barlow (1982: pg 243) has spelled out the fundamentals of one such scheme. He writes:

> "If our use is to be accurately balanced, four things at least are needed. Firstly, we need to get adequate information from our muscles (and from the other parts concerned with movement). Secondly, we need to receive this information accurately with our brains without obscuring it. Thirdly, we need to activate our muscles so that they do what we want, with a minimum of misuse and fourthly, we need to know how to come back to, and maintain, a balanced resting use of our bodies which will interfere least with our functioning."

How would you assess this?

DG: Well, in one sense you can't argue with that. It's true we need information from our system which is as reliable as possible and we need to respond to it with a minimum of misuse. We all want to carry out our activities without getting in our own way, but the third point coupled with the fourth can be quite a source of confusion. What does it mean to come back to a balanced resting use of our bodies? Come back from what? From the active use of our muscles? From active misuse? In order to maintain this resting use which interferes least with our functioning? But how can we be functioning and resting at the same time? On the face of it, it would be easy to be confused. I'm not sure what Dr. Barlow meant here in his own thinking, but such a statement can be interpreted different ways with quite different results. Are we trying to get to and stay in some ideal Alexander posture in between or during activities or are we looking to undo interference in the middle of any activity?

My belief, backed up by all my research in anatomy and physiology, is that we have a system within ourselves which allows us to recognise and be sensitive to misuse or interference at anytime in the middle of anything. And we have built into us the means to respond to our sensitivity in such a way as to let go of the interference in the present moment in the middle of any activity.

So I'm not convinced of the value of the idea that we have a "resting" or home-place to come back to, if indeed this is what Barlow is referring to. First of all, it seems to define any movement that we do as something that starts from this "home-place" and presumably ends up back there. I find that such phrases often reinforce people's existing ideas that there's an ideal, balanced resting place in which, quite naturally, they will want to spend as much time as possible.

But in fact, there is no individual ideal in that sense. The ideal of a postural norm, scientifically-speaking, is not much more than an artefact of measuring and then averaging the state of many people's habitual postural misuses. For example, if someone habitually is tight and arched in his back with narrowing and pulling back of his shoulders and he then undid those interferences, where would he end up? Where will someone go when he stops trying to get himself somewhere? He will go to his particular non-interfered state of relationships at that moment in that particular activity given his particular structure.

If the person then turned to the right and picked something up without interfering, however he did it would in a real sense be the norm for that particular movement for that particular person. Each of us has the possibility of accessing an enormous range of non-interfered movements using the bending and rotating motions in the joints of our legs, arms and torso just by stopping interfering. Hopefully, you'll extend that repertoire and range as your skill increases and as your body opens up in response to a better use.

You may well discover then as you come out of your habitual misuses you will find yourself consistently in a number of recurring postural organizations. For instance, in normal standing and walking, being relatively upright is functionally good for more activities than being bent over to one side. But it doesn't necessarily mean that it's a posture you should direct for or aim to always return to after "activity". After all you didn't have to define it in the beginning, you just ended up there after undoing the interferences. And you won't have to stay there, you'll be there as long as you don't interfere and as long as that particular upright activity carries on. And over time, as your system adapts and changes you'll probably end up in ever-changing and hopefully better organizations.

So what I'm saying is that it is all too easy to carry our old end-gaining habits of thought over into the paradigm of an ideal we are striving to achieve where the attention is more on the use as a goal than what we are using ourselves for. We can change that emphasis to where we simply aim to free ourselves from our restricting habits off into whatever activity we want to do and once we've accomplished that, we'll go off into the next thing and then the next thing. In other words, getting out of stuckness rather than trying to get somewhere or get back to somewhere.

SC: It's not positional then?

DG: No, it's certainly not positional. There may be better or poorer use but there's not right or wrong positions. Bending your spine will be the most appropriate response in some movements and not in others and will happen automatically unless we are holding ourselves from being free to bend, either physically or in our fixed concepts. This is, of course, why we have the flexibility and range of our joints: to be able to use them in various movements. It's just as much of an interference to hold one's torso in a "right" position and think one should never bend as it is to collapse or pull down (though maybe not as damaging).

We have to realise that our use in a physical sense is inextricable from our whole conceptual framework — our patterns of thinking and belief and that we can't expect to change the way we use ourselves without changing all our ideas. This means that we can't use traditional conceptual "hooks" like balance, weight, alignment, gravity, and so on without limiting our ability to come to a new use. We have to give them up because they inevitably have associations and connotations for people. For example, the very idea of weight is something that implies a whole range of other phenomena that will inevitably draw you back into the habits you'd just changed.

In fact, the hallmark experience of this change to better, more whole functioning is that you don't have weight, you have lightness. "Weight" as a concept no longer has any value or relevance in this experience. But instead of letting go of the concept of weight we hold onto it by trying to explain, for instance, that the weight was released into the ground when we released up. But of course a set of scales isn't going to register any difference. To entertain the thought of how weight fits in and where it goes is to weight ourselves again — which drags us back to the same old use.

SC: When you're teaching you're careful that the conversation doesn't wander onto the weather, for example or other topics of everyday conversation. You're trying to keep attention on the experience of the person and tapping that first-person experience in order that they might learn something about their use/misuse.

DG: Yes. If someone comes to me to learn how they interfere with themselves and how they can undo that interference so as to come to a natural and integrated functioning, then what's the point of spending the time paying attention to other mundane things immaterial to the job at hand? All we have when we use ourselves is our own consciousness. We certainly can't use our teacher's consciousness when we're out on the street alone and we're not going to get far with the faith that good use will happen automatically simply because we attended lessons. It's a learning process and what someone is learning, at least in the beginning, is to find out what they do to interfere, how their whole way of feeling, responding and reacting locked them in to their habits, and how they can respond constructively to what they notice.

SC: So how did you come to this approach?

DG: In one way I don't feel that I personally came to it. It's what I always thought the Technique was about from reading Alexander's books after having my first lesson and it's what I thought I was being taught by my teachers and in my training. I thought it's just the nature of the principles that Alexander discovered. Could anyone really argue with the truth that we're trying to find out what we're doing that isn't so constructive and then undo it?

With a bit more experience now, and lots of teaching in all kinds of Alexander schools, I can see that this is not how lots of teachers view the work at all. And many of my own teaching experiences and insights went into the evolution of how I work now.

Nevertheless, if we're talking about an approach of undoing interference — a coming out of whatever we were doing to interfere with our innate harmony — rather than finding a right doing to replace a wrong doing, then we must obviously become aware of how we might be interfering so that we can undo it. If you're not aware that your shoulders are held up, you cannot even question why they are held up. You can't stop doing something unless you know you're doing it. So part of our job is to help people become sensitive to interference and understand it well enough that the means to undo it is clear.

Initially, the particular way of working I've developed did come out of looking carefully at what happened in my own teaching. For example, someone comes to me who has a shortening down her back that creates an arch resulting in the upper part of her back and head and neck being pulled back and down. Then, with my help, she manages to release out of that pull into more length and more freedom. There's obviously been a change which both I and the pupil notice. I, the teacher, would see that change as a positive one — a lengthening and opening of the pupil out of her habit into more uprightness. But so often the pupil would experience this change for a moment but then, almost immediately, try to get out of it and back to where she was before.

Noticing this happen so many times, I got curious what the person was going through. Why wouldn't they happily allow themselves to stay somewhere that was so obviously better (to me)? So I decided to ask when I noticed someone go into that release and opening and then start to tighten back to what he or she was used to, "What are you feeling right now that you're wanting to get out of? And what are you looking for when you get out of it? Can you even feel that you're not willing to stay where you've just allowed yourself to go?"

It was surprisingly common that when there had been a positive change in a pupil's use from my point of view the person would not feel the change as good at all. Maybe she felt bent, she felt uncomfortable, she felt unstable and precarious, or even, "somehow not me". And, of course, unless I can then find out from that person what it is that she's experiencing I don't know that her experience isn't matching my experience. For me it might be a wonderful change but for the student it might be one of the most strange and uncomfortable things that's happened to her recently and she's going to get out of there as fast as she can.

There are a lot of teaching opportunities in those moments. When I started to step in and ask the person to tolerate her experience long enough to describe it, then one of the things she's having to do is to learn to inhibit a reaction against the experience. And second, as she describes it she gets to recognize just how powerful the feeling is and how she'd normally react to it, and lastly she will find out, as we examine the situation together, that her sense of what's happening is not what's actually happening.

For instance, she looks in the mirror and she finds that she does not look like what she feel like — she's not actually bent forward and hunched compared to "normal" — but instead she's much more upright than "normal".

SC: So you think the mirror is an important part of the learning process?

DG: Yes, it certainly can be a great way to help the pupil call into question those experiences that normally would never be questioned. Take, for example, that very common "good posture" habit, "Stand up straight, chest up, shoulders back" where the back muscles end up tightened down so the lower back is more arched than normal and more narrowed with the chest held up and shoulders held back. Suppose, with your help, this pupil had just released up out of that pattern so that he is now more upright, there's now more width and more length in his back, his chest is more free for breathing and his shoulders have been released from being so tight and have come a bit more to the sides rather than be held pulled backwards.

It's useful to use the mirror so that the person can compare his feeling of this change with what's actually happening. So I might say, "How do you feel at this moment? Did you notice a change?" And they reply, "Yes, it feels different". Then I ask him to elaborate, "Different how?". This gives him a chance to get in touch with his own experience and to put it into his own words that make sense to him. As he articulates it to me, it gives me information about how he is interpreting or misinterpreting that experience.

This is different than the mode of teaching where the teacher remarks on changes to the pupil saying what's happening from the teacher's point of view — "Ah, that's good, did you feel that release and lengthening?". In that way of teaching, you can often see pupils searching around in their feelings because the teacher noticed some change so they'd better find it too in the same way that teacher feels it. Often the pupil doesn't even have the ability to feel what the teacher feels or sees, and if they do they almost certainly don't feel it the way the teacher does. This leads to confusion for the pupil, and a tendency to try to notice which leads to even more narrowing to feel the unfeelable. But if I ask them an open question — "Did you notice anything, and if so, what?"; if I can get them just to report on their first impressions in their own words without "processing", it can be very revealing.

SC: So do you cue into that and say that they don't have to come up with something positive as it were?

DG: Well, I think one of the best places to start is just to ask the non-leading question, "Do you notice a change?" In the case I was using above the pupil reported "I feel much more forward" or "I feel bent". And then one can draw their attention to the fact that they've mentioned a positional description — a change of position or shape. I would then ask if they noticed any qualitative changes.

SC: Before you go on to that. Why is the positional element the first thing that comes to mind?

DG: Well, actually, more often, the first thing a pupil will say is that it feels "different". And, that's accurate; I'm sure the pupil does feel very different. But that only carries one bit of information —that something has changed. Whereas, what I want to get at is what exactly changed, not from my point of view, but from the pupil's point of view.

That's when the pupil usually describes how they feel in such a different position. To me it simply shows that people are coming from a way of use where they feel themselves by and large in positional ways. So that's what they feel first; the postural shape they are in. What does "posture" imply to most people but position, hence when they look at their posture they look at positional information. For most people there's "good" posture and "bad" posture, therefore positional sense is very important for them. But when you start asking people what they're experiencing and you start to see that there's a common pattern in what people report — it's not really a question of why, but here it is.

In fact, a lot of people don't normally sense the qualitative stuff at all. Many people are quite happy to go to a position that feels right positionally even though it's very tight quality-wise. In short, they'd much prefer to be right and tight than feeling free and wrong or bent. Everyone gives lip-service to the idea of freedom but when it comes to the experience of it — when it's actually there in front of them — they don't want it at all. That is, they don't recognize that odd thing they are feeling as freedom. I must say that a lot of times people do notice the qualitative aspects, the freedom and the lightness and love it. I am only talking here of those times when the change feels so unfamiliar as to seem unacceptable.

When this reaction to the odd feeling happens it's important to get somebody to tolerate the change and its accompanying strangeness long enough in order to bring in their discrimination, not least because one of the things that's unreliable in terms of sensory appreciation is that people are looking at inappropriate aspects of their experience.

Let me be clear here, I'm not saying there's anything unreliable about the senses, rather it is the appreciation or misinterpretations of the feeling that puts us wrong. And we need to learn to change that because accurately interpreting our experience is extremely important, essential in fact, if we're to use ourselves well.

This seems to me to be the essence of the work. It's a new skill that we're developing rather than a question of going back to something we've lost. We're going on to develop something we've never had which is a conscious understanding of how to use ourselves as opposed to the unconscious integration of someone who just had everything working well from the beginning. As a consciously-defined skill one needs as much information as one can get and to make sure that the information is, literally, as reliable as possible. Of course, our system is incredibly sensitive to what is happening. It's just that we misinterpret it. And we misinterpret it largely because of what we're looking for and what we thought was important.

Let's go back to our example. The pupil in the lesson felt herself change in a way that is very different to her from normal. When I ask, "Different how?", she says, "Much more forward". Then it's easy for me to ask, "Well, is that all you're feeling? Did you feel any quality changes?". By qualities I mean looser *versus* tighter, lighter *versus* heavier, more breathing *versus* less breathing, more aliveness or presentness *versus* less aliveness or presentness, more movement *versus* less movement, more alertness *versus* less alertness, more groundedness *versus* less groundedness, and so on. Now, of course, she may not feel any one or every one of those but more often than not she'll now recall some qualitative experience that goes with her changed positional sense, but which she wasn't even noticing this until I asked her specifically.

In other words, the feeling of being bent forward was so strong that it dominated the experience and set off all the red lights and sirens saying: "Bent! Bent! Bent! Straighten up! Straighten up! Phew! Thank God I've got out of that. Now where's that freedom that I wanted?" So that's where one needs to help the pupil inhibit their reaction to the feeling long enough to allow all the information to come through and be absorbed.

Let's say, for instance, that you, the teacher, notices that she has come up and lengthened and widened in her back, her shoulders have released and come into width too from being held back, and she is breathing more.. You'd be expecting her to say that she feels lighter, more open in the back and/or that her breathing is easier. But this is not what she notices. Often the first thing she'll say is,, "Not only do I feel bent forward, but my shoulders are rounded too".

So I'd then say, "OK, let's collect all these bits together" and I'll just stop the hands-on lesson at that point and say, "Well, I think we need to talk about this to see if we can make some sense of all of it. You've experienced it, but what does it mean? Before you throw it out let's see if there's some significance to it".

If I just take a few minutes and point out that the feeling of "more forward and rounded" is a positional description, and then ask her about any quality feelings, especially ones in contrast to her normal state before.. Only then will she likely notice one or more of the "lighter, looser and more breathing" feelings. With the aid of the mirror I can help her understand that she is not actually bent forward or rounded in the shoulders.

She feels it as: "I *am* more rounded and bent forward". She's discounting the quality feelings and saying, "Yes, I do feel lighter, but" — there's a big but in there — "but, I'm so bent forward". In other words she's saying "Yes, I've got some of those things but I don't like where I've got them. I don't want them there, I want them back in the position which I'm used to". The "but" means, "Yeah, I've got them but who wants them here? Let me go back to where I'm used to and see if I can get them there".

Also very interesting is that someone will say, "Well, I feel I'm like this" and to show me she'll do some totally exaggerated bending that bends her about 2 or 3 inches further forward and actually makes her look like a bent little old woman. The important thing here is that sometimes the pupil has to do the movement she feels in order to show me what it feels like to her. The exaggerated thing they show me is not what's happening, and you'd think that the pupil would know this because she's the one who had to do this whole other movement to show it, but the fact is that most pupils don't quite appreciate the significance of what they're doing in those moments.

It's a very strange thing when you think about it. It's really an analogy. They're saying in effect: "I feel *like* this. I feel *as if* I'm an ape. I feel *like* I'm bending forward. Let me show you what I mean". So it's an odd thing that someone has to do something different in order to show me what it feels like. If pupils took in the significance of this they would recognize that it is not quite what they feel it as.

Another thing I want to help the pupil learn is that what she's feeling is not an absolute thing but a relative thing. She's not feeling the position she thinks she is feeling, that is, where she ended up; she is feeling the change that just took place. She's not feeling qualities like light, loose, free, etc. She's feeling light-er (vs heavi-er), loos-er (vs tight-er), more breathing (vs less breathing), i.e. how she has changed — she's just lightened, opened and started breathing more (as opposed to tightened, got heavier and restricted her breathing).

There's more information too. She's feeling the direction of the change. In essence, her senses are telling her directly that this lightening, loosening and breathing happened in a more forward direction. *Which is totally true information.*

Now, it may be interesting to know the positive change that took place, but it's not relevant in any practical way since she'll didn't directly do that change, nor will she want to try to recreate it. After all, it's already happening automatically for her. It's not important where she's got to, but what is important is that she can look at it the other way — where she's come from before she changed. That is to say, if she is now more forward than she was a moment ago, then, a moment ago, she was more back. If she is now lighter, then a moment ago she was heavier or pulled down. If she's now more open, then a moment ago she was more closed or narrowed. If she's now breathing more, then a moment ago she was breathing less, and so on.

In a literal sense, a moment ago she was "heavying" (being a sort of down-ish direction instead of "lightening" which is an upwards direction), "closing" and "stopping her breathing" backwards. That is, she can now begin to take in what her feelings are really telling her — that she was closing herself in a heavy-ish, down-ish, backwards direction, which is also totally true information. And very important information at that, since that is her habit and precisely the interference that she needs to stop.

That's a reliable appreciation of the feeling.

It was all really there in the experience from the beginning though she was misinterpreting the feeling (unreliable sensory appreciation). If she can put it together accurately; if she can change her point of view this way, then what her own senses are telling her directly is not unreliable in the slightest. Her own senses are telling her exactly what she was doing a moment ago and what her habit is.

Thus, if the job at hand is to finding out what our habits are so we can undo them, then here right in front of us is what we need to know to accomplish that. However, this is only possible for someone who is in touch with their experiences accurately. For people who have been stuck in habits, out of touch and misinterpreting their experience day after day that can take a while. That is in the nature of any learning process, but at least they will get there quickest if they are taking the most direct route.

There are lots of ways to help people uncover these misperceptions and see through them, but using a mirror can help out in interesting ways. For instance, to use our example again, if a pupil has just released out of a heavying, pulled down, tightening state into more lightness, looseness, and length, you can ask her to do something that sounds easy, but is actually very difficult. While she is in her new released sort of organization, you can ask her to *stay looking in the mirror* and then, very quickly without thinking about it, go back to what feels normal — in this case asking her to go from what feels bent to what feels straight.

The important part of this exercise is that the pupil actually gets to see in the mirror the change from one use to another. It's most instructive to begin this right from the first lesson. What's important is that the person keeps looking in the mirror, and I repeat this because this is quite hard for most people to do. I'd say at least 9 out of 10 people will immediately turn away when I ask them to go to what feels normal. They will turn away from the mirror and sort of gaze inwards, as it were, to feel out what feels normal. Since they aren't looking they won't see it happen in the mirror.

I normally stop it at that point and point out to them that they may very well have felt the change but they didn't see it happening. It appears difficult for people to stay connected to what's outside — in this case watching themselves — and observe the process of change. The draw to narrow in to "feel" is too strong.

If this happens I just do it again after spending a moment getting them to come out of their habit again and make sure that this time they do stay looking in the mirror. Then I get them to describe what they saw.

This step opens up another important bit of learning. The pupil feels herself forward and down in a bend and she might describe what hse did to get back to what feel normal as, "Well, I opened my chest up so I'm not so rounded". She's obviously describing this from her point of view in the way she felt it and saw it. I can then point out that yes, she does see it that way and that is true from that point of view; her chest did go up and probably feels to her as more open since the front seems more expanded now. This expanded feeling is helped out by the fact that she also pulled her shoulders back making her chest seem bigger in front.

I have to remind her to check the qualitative feelings so that she can sense that while her chest feels more expanded it is also more held and fixed and her breathing is not as free as a moment ago. However, if she remains within her habitual point of view (her habitual appreciation of her sensory experience) she will be caught in a dilemma. Either "open" her chest up and be held, or let it "down" and be bent and rounded, though maybe more free. Neither option is very appealing, and you can see why this point of view tends to trap the pupil into the habit — the alternative feels worse.

It is now my job to help her reinterpret what she saw in a way that will not only give her a different way to respond, but which is actually more accurate (a reliable appreciation). It's not that she opened her chest up. That is only the end result of an actual muscular tightening down of the muscles in her back that levered her chest up in the front. This muscular tightening down not only stiffens her spine, but also holds her ribs fixed at the back which is why her breathing is not actually free and open even if her chest feels bigger. I might also explain how she's also not free anymore because of pulling her shoulders back which tightens lots of big muscles that surround her chest and further restrict its movement.

Habitually, she felt her chest come up — a "good" thing to her after that "horrible forward bend" I took her into. She did not feel it as a tightening back and down in her back. But, after I have explained this and showed her with my hands how the pull down in back raises the chest in front, we can do the "looking in the mirror thing" again where she goes back to her habit (what feel "normal").

Usually, this time most people are not only starting to see the change back to habit as a pull down in back and hoisting up of the chest, but even better, they are actually starting to feel it that way

What I'm getting at is that there is an anatomy to habits as well as an anatomy of physical structure. To reveal this the teacher needs to help the pupil to bring to consciousness and express what she is feeling so she shows herself how she feels things. Of course, it is also essential for the teacher to find out what the person is experiencing. Only if I know how the person is interpreting her experience can I know exactly how she is misinterpreting that experience, and only then can I help her see through that misunderstanding and come to a more accurate and reliable appreciation of her experience.

I've gone on a bit here, but you asked how I came to this teaching approach. It's been my experience that when I bring in this exploratory dialog right away to help pupils get more in touch with their experiences and more accurate about what they are feeling, they come to a more reliable sensory appreciation of their daily use quite quickly and so their overall use improves easily and quickly too.

Years ago in my teaching, if I just "gave" them the changed experience and expected them to feel it and understand it in the same way I did, they got all tangled up in trying to keep the feelings and trying to get somewhere "right". Most people could "have" such changed experiences at a teacher's hands in lessons until the cows come home, but not really change that much on their own in daily life because they never understood what they were experiencing. In fact, they usually misunderstood the lovely experiences they were having. That's when I realized that it is not the experiences themselves, but the understanding of the experiences that is key to learning. This is where there is no way around taking the time to find out how the pupil is interpreting their feelings and what they are trying to do with because of their interpretations. A big change in my teaching I can tell you!

SC: That's interesting. You are really getting into what the pupil thinks and feels. However, I'm sure a lot of Alexander teachers would be reluctant to go there, especially if the territory of feeling gets into anything emotional.

DG: Let me clarify here that this getting into what the pupil thinks and feels is not for mme to find it out. My job is to help the pupil get clearer about what they are already thinking and feeling so that they can be more in touch and more able to question their interpretations of their experiences.

But yes, I think you are right about the reluctance of a lot of teachers to go there. There has been a general tradition that working with a pupil's use means their physical use only. However, the model of communication that I just outlined, and the learning that comes from it, is an extremely fascinating process. The so-called emotional things that come up, if we understand them, are not anything to be frightened of. In fact there is just as much learning to be had there too.

I'll give you another example that deals with an emotional feeling. When a pupil lets go of holding patterns that have been so long-standing that they have become built into the person's identity, the person often end up feeling very vulnerable — they experience what we would sense as an openness but to them it feels very vulnerable, very raw and exposed. Normally, they would react emotionally to this vulnerable feeling by rushing to close off again, to feel "safe".

In the lesson, by taking the time to look at the feeling and the person's interpretations and reactions, I can help them see that this vulnerable-but-open feeling reveals a great deal about the way he has been operating (held and "protected"), and how the feeling of "vulnerable" is not a description of what is actually happening to him in the moment, but the misinterpreted feeling of a change from the moment before. If he wants to keep that openness and expand into that new space in his life, he will have to deal with and live the openness without shutting it down.

For example, if he expressed the fact that he feels vulnerable in a lesson after opening up and having his breathing become more free, I'd help him see that what he's been doing is hardening and closing himself in a way that makes him *feel* less vulnerable. The only thing that changed right now was a bit of a release and the freedom of his breathing. The world hasn't changed so that there are suddenly more people attacking him. He's simply breathing more freely and feeling more. His faulty appreciation was that the feeling of being more held equalled being more protected and that the feeling of being more open equals actually being more vulnerable. The accurate reality is that the change is not about any actual territory of protected safety or vulnerability, but simply and only about holding or openness.

In fact, a moment's thought will show that, if anything, it is the opposite — being more held doesn't make one more protected; it makes one less aware, less available for response and therefore probably more vulnerable.

The other big thing he can learn from these experiences is that he has a habit of inter-preting feelings as it they were a description of the state he is in — "I *am* vulnerable" or "I *am* free". In reality these feelings are not a state to be in, they are a *response*. It's not, "I feel vulnerable" (full stop). It's more of a question: "I feel vulnerable *about what?*" or "...vulnerable *to what?*". That is, what you are feeling is the feeling of a response *to something*.

In the example I just mentioned there can be no sensible answer to that question since the outer world did not change at all when the pupil released so there is nothing new to be vulnerable about. But unless the pupil is aware of what his feelings are feelings of, he will simply react to the feeling itself and, in this case, try to get back to that feeling of being protected which, to him, is the positive side of his habit. The negative or unconstruc-tive side of it, which he will normally not be aware of at all until we explore it, is that he would get that feeling of protection by stopping all the openness of his breathing and the responsiveness of his freedom of movement.

Once he understands this, it's then possible for him to realise that he can relate to a different and more accurate aspect of himself than normal. He feels more vulnerable in the terms of his habit, but that's inaccurate. What he's actually feeling is that he's more open in relation to the world around him, which in fact is quite a good thing. Then he'll recognise that in these new, more accurate, terms he likes that openness, but in the terms he was feeling it before he didn't want any part of it.

He will begin to realise that openness feels like vulnerability in the beginning from the habit's point of view. But when he can inhibit his reaction, tolerate the feeling for a moment, he'll see that nothing has really changed except that he's much more spontaneous, aware and able to respond to situations. Interestingly, this will also contribute to him feeling less of a need for protection, so it's all a virtuous circle. He will have come from a faulty appreciation of his experience to a more reliable one.

SC: Are there any other expressions of openness that commonly come up for people — or do you think that vulnerability is the main element in all of this.

DG: Oh, no, there's lots of different ones. Vulnerability is only one of the things that happens to any one particular person at any particular time. But I don't think one can say that vulnerability is the first or even most common expression of openness and freedom. It depends on the person and where they've come from. After all, what I'm getting at is that perceptions and responses are always in that individual's terms.

Another common one is that freedom often feels like precariousness or wobbliness if a person has been used to stabilising themselves by tightening. This letting go can feel quite unstable even though the openness and freedom actually represents a deeper level of stability than a stability generated by fixity. Another consequence of a release into openness is that some people will experience a greater degree of dimensionality — they feel more present — which can create for some a feeling of discomfort because it's too direct an experience compared to what they are used to.

Often too, there's the emotional magnitude of having just released some major pattern of holding and having a realisation wash over them of how horrendously much they've been doing to themselves for so many years. You can often see the tears come up in someone's eyes — part relief and part sadness — because they have realised the sheer waste of their energies.

Of course, sometimes people just feel much more joyful, free, and light, and usually feelings like those are easily accepted and expressed. But for others who are used to controlling and who do not trust themselves, the urge can be more to suppress the exuberance and rush of spontaneous feelings. Freedom can be quite difficult and uncomfortable to suddenly experience, like a caged animal suddenly let free from their small but known world. People are often wary and unwilling to venture out into the immensity and the unknown. At least at first… and at least until they more fully understand their experiences. Again it is the understanding that makes all the difference.

To really have it work for you, you cannot just sit back after a teacher has facilitated a change and groove on the feeling. You have to allow yourself to expand in new and uncontrolled territory. For instance, when you let go into a more open use in your torso area for example and experience release in your breathing, in a larger sense it's not that your breathing has become more free, it's that you've opened more to the room around you. You are more out in the room, and the room and its air is more in you. But that kind of freedom can't happen and can't be sustained unless you move right out into it and willingly occupy it. If you've actually released out of something and you are more free, then the last thing you want to do is to try and hold onto it and stay in the place where you first got it because you then close it back down into me in here and it out there. After all, it is not just a new use inside yourself you really want, but a new use of yourself in the world…

84

SC: Yes, didn't Alexander used to say, "Now you've got it, just throw it away."

DG: I've heard that one, and if he was referring to the feelings after a change, it's certainly something I'd agree with. Because that's the next step. If someone can release out of their misuse and out to the wholeness of themselves then that's only part of the process. One has to then adopt this whole new self and live it out into the world or it will just evaporate away again into the old habit.

It is easy to see when someone has stopped short of the second part of the process. Yes, they have been able to allow a release into themselves, their eyes may be open but they're not actually connecting to the world around them — all their attention is turned inwards. They've stopped that release at the edge of themselves by trying to stay aware of themselves and how different or how nice the changes are, rather than let the release flow right out into the room and into activity. Of course that's the old habit — to limit everything to yourself, to try to get a good feeling and then carry it around with you. This just gets in your own way. You've released out to the whole of yourself but have not gone the rest of the way to release yourself to what's around you, into response to the world and further changed feelings.

SC: Questions or problems of identity are obviously important here.

DG: Oh yes, and it can be quite scary stuff for someone used to controlling and hanging on to themselves and defining themselves by the familiarity of their habits. The more people come out of this, the more they gain that quality of presentness which is partly being expanded right into the world, but also partly being anchored to a fixed self. If I release into myself and beyond into the whole space around me I can't really say what's me and what isn't, where I end and where the rest of that starts. If I'm open enough through myself to perceive the ground then it's part of my experience and, therefore, part of me. It's not a separate thing.

SC: I think to a lot of people that will seem dangerously mystical.

DG: Ha ha... I'm sure, but there's nothing mystical about it from my point of view. It's a completely tangible experience after all. It is not a kind of goal of what should be, or what we are aiming for to transcend this reality. It is simply what does happen. And it happens regularly enough, and naturally enough, when people do allow themselves to come to a very different use — a different use which they themselves, in their own words, describe in terms of qualities of more presentness, feeling more a part of everything, more expansion in space, everything more three-dimensional, sharper and richer in colour, and so on. It happens this way consistently when we stop trying to control things so it must be an innate aspect of your nature not some transcendent state on the next level that we have to try to achieve.

I don't think there's anything mystical about it, though it is undoubtedly a very profound change in the most literal sense. I am simply talking about our natural expansiveness to, and inseparability from, the world around us. Unfortunately, too many not only have lost that expansiveness but constantly lock themselves away from it by pulling back into themselves trying to observe and control themselves — they narrow in to try to get their bodies to expand outwards — instead of trusting their own system enough to let themselves flow with the direction of that expansion out into the world of activity and response.

Come to think of it, I can see why for some people that does qualify as somewhat mystical after all...

Chapter Four

On Fitness

Extracts from a Conversation
with DAVID GORMAN

Interviewed by Sean Carey in summer 1988

Reprinted from *The Alexander Review, Vol. 4, 1989,*
Subsequent editing, June 1993

Key: **SC** = Sean Carey   **DG** = David Gorman

SC: A lot of people nowadays are into fitness regimes of one sort or another. Clearly, people's motivations in taking up an activity vary, but a fairly common one is the attempt to achieve a high level of cardiovascular fitness. Very often sport is seen as the perfect antidote to the sedentary job. How do you see this fitting in with our work?

DG: Well, people are in jobs that they feel aren't very physical and which don't demand very much of them other than their mind. And very often there's a fear factor in there somewhere: "I'm going to have a heart attack or get a bad back unless I do something very active and get my heart rate up and strengthen myself". Another element is that because of all this "unfitness" they want to make up for lost time quickly in their working out which leads them easily to the "no pain, no gain" syndrome: that if the activity isn't strenuous it's not building up any worthwhile strength and if it isn't hurting it's not doing any good at all.

But the big question is: what are we trying to achieve by this so-called "strengthening"? Obviously if someone is not working on changing their habits, all they end up achieving by exercising their habit is to reinforce the vicious circle whereby they get stronger at their habit. In other words, if they're pulling down and tightening in ordinary activities, they'll just pull down and tighten that much more when they exercise. And even if they do achieve a higher level of cardiovascular fitness and they've become stronger at contracting, bracing and tightening with weights or fitness machines — the type of activity that makes somebody hard and firm — then they will actually need that cardiovascular stamina in order for the heart to be able to push the blood through those tightened and braced muscles. In fact, it's very revealing that often when someone like that manages to achieve the sort of global release in a lesson that takes a lot of pressure off both the contents of the torso and the musculature, their blood pressure can drop so radically that they'll see black spots in front of their eyes, feel light-headed and maybe even pass out. They're no longer so hard and tight and hence the blood can flow through rather than being forced through the veins and arteries that had previously been squeezed in the muscles. It shows very directly how much the heart had to work when someone is maintaining a strongly held musculature. Previous to this experience all they may have felt was that exuberant and seductive exercise after-glow.

SC: There are now signs that the more thoughtful writers of sporting literature are moving away from the simple idea that cardiovascular fitness equals health. The argument is clearly put in, say, Dr. Kenneth Cooper's book *Running Without Fear*, (Bantam Books, 1986) which is an attempt to counter the criticisms and the fears that accompanied the death of the

USA's running guru Jim Fixx (author of *The Complete Book of Running*, Random House, 1977) after a training run in 1984. He states very clearly that people should not over-exert themselves, have regular medical check-ups and that it is a myth that the more we exercise, the healthier we are. And yet sensible though all the advice is, Cooper doesn't really consider the fundamental thing — our use, except indirectly insofar as certain conditions like heart-attacks often express gross misuse of ourselves.

DG: Yes, it may be better that people don't strain themselves so much in whatever exercises they may do, but with or without strain the brute fact is that without changing their use they haven't changed the foundation, they're just practicing their habits with a little more strength and vigour. If they haven't changed their use, they haven't changed any of the neuromuscular patterning built up over the years from all their previous use. They're still deepening those neuromuscular connections, deepening the established arrangement and organization of the muscles, the connective tissue and the joints and, of course, deepening their familiarity and adaptation to the sensory side of their experience. The territory we should be exploring is not so much: "How do I change this movement so I can do it better?" but "How do I change the "I" that is generating the movement?" The movement will then, of necessity, change in accordance with how I change. In other words, a certain moderation is probably a good thing, but you're quite right that without changing the foundations there's always a flaw built into any system no matter how benevolent the consequences.

SC: Still a lot of Alexander teachers go to the opposite extreme and actively discourage their students from performing almost any activity whether it's running, rowing or weight-training. Obviously that's problematic to people who say: "Well, what do I do instead? I want to be involved in something". That seems to me to be a perfectly valid question.

DG: Clearly if someone loves doing something it's obviously to their advantage if they can do it with good use. On the other hand, if they can't bring in good use and they still do it in spite of all that, it's their choice and they must live with the consequences. But it's rarely as clear-cut as this. Let me give you an example: I was giving lessons to a very keen oarsman. It soon became clear that he wasn't making much progress with the Technique because of the sheer amount of time and effort he was putting into rowing which had resulted in a massive pull-down the front and forward bend in his upper back. So I mentioned to him that he had two options. First, he gives up rowing for a while, learns the Technique and later on sees if he can take it back to the rowing; second, he carries on rowing but for the moment foregoes all the speed and effort and spends more time exploring how he is rowing. He wisely chose the second course. I went to work with him at his rowing club on a rowing machine and could observe him at first hand on the river. It was very productive and (not coincidentally) his rowing improved in speed and ease when he did take himself back to competition. But, ultimately, how we worked was his choice. It's certainly not up to the teacher to make the choice. We can only say what we ourselves might do which may or may not have any relevance for the student.

I think that often it's not a question of saying to people that this is the wrong sort of activity and you shouldn't do it, but rather making a different emphasis and working with people in such a way that helps them to develop a sensitivity to their use so that they are able to notice when they're not using themselves well in whatever activity they happen

88

to be doing at the time. If we as teachers actually want people to develop an internal register and value-system for their use, then we have to allow them to make experiments and discover what does and doesn't work for them. The more they develop that register the more they'll be able to assess the value for themselves of any particular activity at any particular place and time.

SC: Yes, I take your point but it's precisely that internal register of use and mis-use that is often omitted from even fairly sophisticated attempts to link particular disciplines with improvements in co-ordination. For example, the anatomist Philip Tobias became interested in certain similarities between ballet and karate. He writes:

> "I managed to obtain the services, and indeed the enthusiastic co-operation of the leading karate-ka in Johannesburg and of a group of ballet dancers. The groups I have worked with enabled me to obtain a greater understanding of the movements of ballet and karate. They helped me to comprehend that to achieve physical skill one must acquire poise. To acquire poise, one must convert malposture to posture — by undoing the torque. One must de-rotate the body." (*The Tottering Biped*, Committee in Postgraduate Medical Education, University of New South Wales, 1982:56)

Tobias seems to be suggesting that performing ballet or karate will of themselves — and in a simple, mechanical way — allow an individual to achieve balance and poise. Now my experience of several forms of karate is that it isn't that simple. Moreover, most karate is taught in a very stiff and held sort of way. The idea of releasing and opening into movement is not widely practiced. In fact, a few weeks ago I was talking to a chap who practices the Shotokan style of karate. He adopted the very low stance that they employ and proceeded to demonstrate several different types of kicks and punches. There was a tremendous amount of tension and effort in all of this, not least because he'd been taught to pull his shoulders right back thus narrowing and shortening his torso and gripping his legs.

DG: Well, I think it's possible to perform karate with good use but it's certainly not going to be done by the person who assumes a position or shape by pulling his shoulders back. As you suggest, there is nothing intrinsic to karate, ballet or t'ai chi that will by itself achieve poise or balance. For that matter, there's nothing intrinsic to the Alexander Technique that will automatically do that either. It's how we go about it that counts. However, I would agree with Tobias that to achieve skill one does need "poise" and a good functioning "posture" otherwise there is always a flaw in the foundation no matter what the activity. What we need to work at, whatever the method we use, is finding how we interfere with our innate poise and then get out of the way. I have no doubt that through almost any method or form one could potentially achieve a deep integration and harmony — look at *Zen and the Art of Archery* — but of course one has only to look around to see how often it actually does occur. While it is certainly possible in more form-oriented pursuits it seems to me that it is precisely this territory of how to get at the malposture/posture, disturbance/ poise, action/reaction, etc. which the Technique attempts to directly address, even though we do not always succeed either. From this point of view, the Alexander Technique is a "pre-technique" through which we gain the foundation skill of coming out of poor use and sustaining better use so that we can take advantage of the technique of our favourite activity whether it be ballet, karate, sport or just plain living.

There is another aspect to your question, though, concerning ballet which is a little more problematic. Ballet emerged, after all, out of a very artificial, stylised court culture in France. The first forms were cultivated to distance the courtiers from the natural, "animal" side of themselves and to make an art form of life! With such a historical background, it's difficult to escape balletic values and assumptions. Ballet is very different, therefore, from the more free-flowing modern forms of western dance or for that matter, traditional Balinese or various African dance types. In fact, the way in which people try to attain the ballet style takes us to the heart of the matter. For the most part, the dancers are taught to operate from a fixed "pull up" like an imaginary line hauled right up through their middle until they identify with it as their badge of "dancerness". With the hours and hours of practice they get very skilled at a controlled holding wherein the most important thing is what it looks like. They are actively engaged, therefore, in ignoring their inner sense of use and concentrate instead on manipulating their sense of shape from the point of view of the visual impact.

SC: So are you saying that ballet is so artificial and necessarily involves so much holding that it's impossible for someone to do it with good use?

DG: I certainly wouldn't go that far because it's possible to work with ballet dancers in such a way that they dispense with some of the tight holdings and forced controlling so they can open up and release into movement. It just can be very difficult, that's all — often more difficult than, say, working with someone involved in a modern form of dance that is more free-form and explorative of flow and integration. But, then again it depends upon the particular person. Some ballet dancers are a complete joy and delight to work with. One can't escape that we are working with individuals no matter what they've got themselves into.

SC: One way out of this dilemma, of course, would be to advise the student that when they're doing ballet they should just get on with it but that the rest of the time, in their ordinary life, they can bring in the Technique.

DG: Well, all of us inevitably do that to some extent anyway. It's impossible to think about the Technique all the time and, furthermore, the more complex the activity, the more difficult it is to employ the Technique. This is why in a lesson you'd normally start with something relatively simple like getting in and out of a chair, walking or picking something up, until people get an idea of the tools they are using so that they can take their new skill into more complex activities. It's going to be much easier for someone to be aware of their use and be able to change it when they're walking down the street than, say, if they're up on a stage performing a violin solo where their living depends upon it. They can't really stop and say: "Hang on a moment, I've just noticed my misuse. Give me a second to work it all out... Okay, now I'm fine. Let's get back into the music".

But any artist has practice time and that's one place they can *begin* to bring the Technique into their work. And if we as teachers can help them with that then so much the better. If a student can spend at least part of their practice time — whether the activity be dance, music, the martial arts, or rowing — exploring their use rather than practicing their existing skills they'll be able to bring the Technique into more and more of their lives including the more stressed real performance. On the other hand, if they find that they can't bring their

90

improved use successfully into the old activity then, perhaps, they have to assess whether they want to express their new self in the old medium. A dancer, for example, experiencing a greater level of integration and flowingness through the Technique might well find themselves drifting towards a different form of dance. But that's something they must decide for themselves in the light of their own experience.

SC: Let me now turn to isometric exercise where, for example, someone puts their hands on either side of a door frame and pushes really hard and yet nothing is moving? How do you assess this?

DG: I must say I'm not really a great fan of isometrics because to put the effort in pushing or pulling against an unyielding resistance is almost always to engage in rampant shortening. It's interesting when you think about it that most of our misuses are isometric contractions. We hold ourselves in fixed positions contracting against our own resistance. Instead of you pushing or pulling against the machine or door frame, you're pushing or pulling against yourself. In fact, you are pulling with one part of you against another pull from another part of you. This doesn't seem to do us much good in our own use, I can't see that it's going to do us much more good doing it against a door. If we can already use ourselves openly without shortening, then there are much better activities in which to practice such a use than against immovable objects. However, having said that, let me re-iterate that I think it really is someone's own choice what they do. My job is to help them discover the full implication of what they're doing and how they're doing it so that they can make an informed choice.

SC: Last time we spoke you told me you had some interesting thoughts on Nautilus machines.

DG: Ah, yes, it's easy for Alexander people to get elitist about aerobics, exercise and the like but it's useful to realize that the equipment in gyms — Nautilus machines being one of the most common — present wonderful opportunities for the exploration of our skill in the use of ourselves. In fact there is an opportunity in almost any activity or situation for bringing to bear these "Alexander tools" if we are willing to drop our prejudices and get down to it.

But to get back to working out on exercise machines… The nautilus machines consist of a range of different apparatus, each designed to work on a specific area — one for the quadriceps, one for the upper chest, one for the hamstrings, etc. The Nautilus name actually comes from a little gearing mechanism shaped like a nautilus shell in the machine which ensures that though the weight you are moving remains constant, the force needed to move it is larger in the middle range of movement when the muscle is capable of more force and diminishes toward the end of the ranges when the muscle's capability diminishes. They generally have a seat or bench to sit or lie on and you push or pull on the handles or levers to lift the weights and work out the specific parts. The weight is variable usually in 10 lb. units starting with zero or 10 pounds and going up toward a hundred pounds or more. You get on the machine and do a cycle of repetitions at your chosen weight load, then move to the next machine and do another cycle, and so on through the whole set of machines until you have "worked out" the whole body.

However, what most people actually do is over-exert themselves by loading up as much weight as they can manage to move through the cycle of repetitions. With that amount of weight they can't actually move it in a free and open way but only by bracing and tightening the rest of themselves. It's easy to tell when a pupil has been to the gym the day before. They come in all hardened, compacted and bound up in themselves from the pulling in on themselves. They think that they're getting stronger and they are — they're getting stronger at compacting themselves and using their energy against themselves in bracing and tightening. From our point of view this is obviously not desirable even though it is a virtue in part of the gym-going culture to have that hard body.

However, there's nothing intrinsically wrong with the machines. If you are willing to change the way you use them you can learn a lot. The first step is to take off *all* the weight and just add on the amount which still leaves you able to do the motion of that machine without any bracing or stiffening. In my experience this will be no more than ten pounds in the beginning for most people. Unfortunately, I haven't had the occasion to work with anyone in the gym enough times to see how far they can go beyond that. In the times which I have helped someone, ten pounds has been sufficient for a very good workout.

Let's take the example of the machine which works the pectoral region. You sit upright and with your arms out to the side, forearms upwards you press in on the apparatus to bring your forearms together in front of you then back again out to the sides and so on. Usually people bear down with all their effort on the squeeze in to the front, typically holding their breath with the effort and then allow the weight to slam them back out to the sides. People working with a trainer will often be told to breathe out during the effort when the arms come together and breathe in when the arms come out to the sides. This further reinforces the impression that the "real work" is in the contraction of the muscles frontwards and the opening out is just to get ready for the next "work out". This is not how the machines are designed and a good trainer will ensure that more equal attention is paid to each direction, yet without care this bias will be reinforced.

But to get back to how to use the machine well. Take off most of the weight. Then the real endeavour is to be able to allow the movements of the arms with the load of the weight without having to tighten through your ribs, abdomen, lower back, shoulders or anywhere else. And without having to interfere with your breathing or force it into a simplistic rhythm with the movement. Let it find its own harmony, its own un-interfered-with rhythm. Can you release onto the seat and use the full range of the machine without squeezing in your chest in front or your shoulder blades in back? Can you support the load during the movement with equal release in both directions of movement, especially the opening-out movement where you must support the load yet still allow the muscles to lengthen? Can you keep the release going with a slower or faster movement? All sorts of productive playing around are possible…

What you will be practising is not bracing and shortening in the rest of you to work out an area through contraction, but being able to remain free and open while doing an activity in spite of its load. This is a very valuable skill and is quite usable outside the gym! In addition, what you will find working in this way is that while you may initially think that 10 pounds is nothing, at the end of those 10 repetitions you will have had as much of a workout as you ever did with 50 pounds. In fact, more of a workout. Everyone I've worked with has said that they feel more of the rush of "exercise" afterwards than they ever did with the heavy weights, and they have the delight of having learned something constructive on top of it.

92

Working this way shifts the emphasis away from the work-out of the part to the use of the whole during any specific activity. The particular machine and how it loads you is simply the medium for the larger task of paying attention to how you use the rest of you. The idea behind it is simple, but it's not easy to do in the average gym environment because you have to spend a lot more time in order to be able to keep releasing throughout the repetitions. This is a good thing from the point of view of the work-out since you then have more time in a released state during the movements. But it is difficult for many people to take the necessary time if they feel the other gym fanatics are waiting to get on the machine and snickering behind their hands: "Look at this wimp, he's only pumping 10 pounds. Hurry up and let us real men get to work!"

Chapter Five

Overview
A Personal Viewpoint

A series of editorial columns
published in The Alexander Review

A World-wide Community
Reprinted from *The Alexander Review, Vol. 1 No. 1, Jan. 1986*

Welcome to the first issue of the Review — a long time in the planning. It comes out of a conviction on the part of David Alexander and myself that there is emerging a growing and active world-wide community of teachers of the Alexander Technique. Not, of course, a community just in the sense of sheer numbers spreading out on the face of the planet all *calling* themselves Alexander teachers but with little communication between them; rather one where teachers from diverse trainings and cultures more and more are seeking each other out to work together, discussing differences and similarities, and learning from each other.

The International Review was conceived as a way to support and focus that constructive energy. Its purpose is to serve this community by linking teachers to each other across geographical distance and to deepen the community by being a forum for teachers (and others) to link to the common core of the Alexander work across any of their stylistic differences. I think it would be in order here to offer some explanation why the two of us were the ones to start this journal (and, incidentally, to give my credentials for this column).

Both of us have a somewhat unique relationship to the Alexander world at large. David Alexander, though not a teacher, has much experience with the work and a deep understanding of the value and implications of the Technique. Through his publishing company he has been responsible for a one-man renaissance of written material about the Technique. In addition to re-publishing Alexander's own books (thereby saving a whole generation of teachers from meeting Alexander only in faint photocopies) he will soon bring out a new edition of Lulie Westfeldt's personal account of Alexander, *F. Matthias Alexander, The Man and his Work* as well as a book about the relevant writings and work of Prof. Raymond Dart, edited and augmented by Alexander teacher and trainer, Alex Murray and the first of a series of anthologies from various writers called *The Alexander Reader*.

Along with his obvious qualifications as publisher of the Review and his talent and experience as an editor (not to mention the fact that he has the same last name) he has another important quality — that he is *NOT* an Alexander teacher and is outside the normal political groupings, hence has a central and widely trusted, but impartial, place in the community.

I have come to this from the opposite direction — from within, outwards. Like all teachers of the Technique I have been greatly influenced by those with whom I trained and like most teachers it took me an additional number of years to find my own style of working which combined my training foundation, my understanding of the principles of the work and my own nature and personality. Because of the work which I've developed over the years on understanding our anatomy and our physiological and conceptual habits, I have been lucky enough to be invited to teach in a majority of the teacher training courses around the world. In my workshops open to the public I've worked with Alexander teachers of nearly every persuasion and countless numbers of their students. That work (and my books) has also taken me into many other disciplines — training schools of bodywork techniques and therapies from massage to osteopathy, performance faculties of colleges, sporting groups, and administrative bodies.

All this has given me a broad experience of the many faces of the Alexander Technique. In the process of building bridges from my work to their practice I've come to see the presence of a common principle underlying the variety of styles, the differences in approach and even the various levels of skill, training and understanding. And I've also learned a lot about the value of each of those differences! Through my courses outside the Alexander world, I've gained a perspective on how other professionals view the Alexander Technique, what they feel they can gain from it and what they see as its weaknesses (not always what I would see from the inside). From many individuals who are studying the Technique with a teacher and came to a workshop of mine to better understand how they are built and function, I've come to see what parts of the teaching of the Technique are not clear to the average student (and, for that matter, to a lot of teachers too). All of this will be material for future columns (and also a series of articles for the Review on understanding our human design and function (this refers to the series, *In Our Own Image*, which is published as Chapter Two of this book).

The response we've been getting from our Contributing Editors and from individuals who contributed articles has been very heartening, as have been the mass of subscriptions that came in unsolicited before the order forms were officially sent out. But so it should be if this periodical is to be a voice for, and a reflection of, an already thriving global community. This sort of professional intercourse is the best thing that could happen to the Alexander Technique after decades of development in relatively isolated cultural or national pockets.

Development of the work has been necessarily and understandably isolated due to several factors. One was the difficulty and expense of international travel up to the late Sixties and early Seventies which restricted the approximation of teachers. Also, before this period (with its apparent fundamental shift in society which resulted in a more global and ecological viewpoint), the culture had not been very fertile ground for the popular growth of such a radically different educational, non-therapeutic technique. Thus, aside from a few major centres, teachers were few and far between. Related to this is the larger factor of the evolution of the Technique itself. We are now, as a collective body of teachers, only beginning to enter the third generation after Alexander and his discovery. Alexander was alive and further developing his work for most of the "life" of the Technique. The first generation of teachers trained by F.M. himself were faced with the dilemma of teaching, each in their own way, what they had learned from him *even while* he was still changing his way of teaching and, further, of realizing that the other teachers who had also learned from Alexander had apparently learned different ways of working!

It was all, of course, new in those days. All of them, like every teacher, were subject to appreciating the Technique through the filters of their own personal nature and experience. None of them had any tradition except Alexander to fall back on. There were no already existing "different styles" so, naturally, pressure was great to teach THE Alexander Technique, and each had their own idea of what that was… I find it not at all surprising that these factors added up to the current diversity of styles. Whatever conflict between them that has developed along the way is equally understandable.

Most of us who are second generation teachers did not know Alexander personally. We grew up, as it were, with the diversity of the Technique all around us. Though it came as a shock to me to arrive from Canada, where at that time there was only one teacher, to the training school in London and discover that everybody didn't teach this stuff the same way, it was readily apparent that these differences had a "history". We had a natural curiosity about these other ways, unfortunately tempered with a natural prejudice in favour of our own way. In spite of all the barriers, though, bridges are now common between like-minded teachers of different styles and levels of experience. And already there is a third generation of teachers appearing who have had the opportunity of gaining the advantages of all this diversity and who want no part of the remnants of past conflicts.

The Alexander Technique has not hit instant stardom in its history but, considering its radical differences from most present-day therapies (techniques, educations, etc.), it's done not too bad at all to have come this far so fast. It's a testament to the transforming nature of the work itself as well as a stroke of luck that the explosive growth of interest in alternate lifestyles, self-awareness and self-actualization in the Western world has coincided with a maturing in the Alexander world capable of supporting such growth. It's a testament to Alexander, most of all, who gave us such a powerful and timeless instrument. It's a testament to those who learned from Alexander, who believed in the work enough to carry it on (sometimes alone into new countries) and who built strongly on Alexander's foundation. And it's a testament to all those, old and young alike, who use this work to transcend their own limits, their own doubts and prejudices and get out there with others who are doing the same thus ensuring that the noble spirit will continue to swell in the future.

Emergence of a Profession

Reprinted from *The Alexander Review, Vol. 1 No. 3, Sept. 1986*

The Alexander Technique has been around for almost one hundred years if we measure from when F.M. began to show others what he had discovered. It is about fifty years since Alexander began to formally train people to teach the technique he had developed. Today the number of teachers is approaching one thousand and there are over three dozen active training courses. We have not stood still since Alexander's death in 1955, for there has been consistent development in conveying the work effectively, with many having contributed a diversity of styles and approaches. Up to this point the Technique has also enjoyed a public perception of quality and integrity. There are certainly no lack of pupils for most experienced teachers and there are far more students wishing to train as teachers than there are training schools to train them. It would seem from the standpoint of duration and development, of sheer numbers and reputation that the Technique has become an established and growing profession.

There is probably no better indication of this than the rousing success of the International Alexander Congress this summer in Long Island. Much of this issue of the Review is taken up with reporting on various aspects and events of the Congress so I have only a few comments. Much credit is due to many people for its success, from the magnificent organization of the whole event by Michael Frederick and the Executive Committee to the inspiring work of the Senior Guest Teachers and the many who gave lectures and presentations. However, I think the real star of the show was the tangible atmosphere of exuberant spirits to which everyone contributed. The very idea of 225 teachers and students all gathered to celebrate being involved in this work and to learn from each other made for a heady brew. I heard many people there speak of "an historic occasion". Now, this excitement is no more than regularly occurs at any professional conference, especially when you get people away from their normal habitat. But this kind of thing has never happened before in the Alexander Technique! And now we have another one planned for the summer of 1988 in England...

Yet all has not always been rosy and in many ways we have not in any global sense really managed to become a unified profession. With the explosion in numbers and the spread of the work geographically has come some disturbing aspects. We may have the existence of a number of valuable ways of working, but on the other hand we also have an increasing isolation and consequent chauvinism of some of these styles. We may have a healthy growth of interest and demand for training schools, but we also have an increasing dilution of standards in terms of available training time and the depth of personal change and fundamental skill that is achieved. In some cases this has resulted in a complete loss of connection to the traditional way of training and teaching. While these conflicts and rifts are disturbing to us within the community, we are fortunate that so far they have not seriously damaged the public standing of the Technique.

Further inability of these factions to communicate and come to terms with each other will threaten to drive some into even more purist stances and others deeper into radical breaks from tradition. But most of all it threatens to further divide the Technique as a profession to the point where we will all be harmed, not just now but for years to come. However, to say that there have been problems does not mean that we are stuck with them. I believe that we all have a responsibility to our profession. A responsibility to help ensure that we escape the fate of some other similar professions who have ended up with several rival and conflicting factions each with their own governing bodies or whose professional training standards can only be said to be in gross disarray.

We can only strengthen ourselves and the work by creating a strong and cohesive profession with consistent and high standards. There is little question that the quality of tomorrow's teacher depends upon the quality of today's training. In the same vein, we can no longer afford to work in isolation from each other without weakening ourselves (by the measure of what we have not gained from other teachers), weakening other teachers (by the measure of what we have not given of ourselves to them), and making it harder for our students (by the degree to which they are kept from exposure to other teachers).

One of the powerful products of this work is the growing ability for each individual to get in touch with his own life and to uncover and develop his own potential to whatever degree of refinement is desired. With much study we are able to begin to gradually uncover our own interferences and improve our own use. But, anyone who also continues to take lessons or do exchanges with other teachers knows how much more someone else can show them than they themselves can find. So to me there is another, and perhaps even greater, opportunity the Technique gives us.

That is the possibility of allowing others to get close enough to us to show us what we are doing and to get close enough to them in turn to show them what they are doing.

To succeed in getting in touch with others like this there must be mutual acceptance and a two-way openness. In this day and age we can use all of that we can learn and anything which facilitates this learning and interaction among teachers is good for the profession let alone the teachers.

The Congress (and future ones like it) certainly came under this heading. But several events occurred just prior to the Congress which also promise to do just that. One has to do with changes in the *Society of Teachers of the Alexander Technique* (STAT) based in London, the largest professional body of teachers, whose membership constitutes over two-thirds of the teachers in the world. The other concerns the formation of a new society in America, where over a quarter of the world's teachers reside. Both of these events will have the result of more closely linking teachers and of including other teachers previously outside any of the professional organizations.

On July 19th at its Annual Conference in London, STAT amended its Constitution (on a motion proposed by myself) to allow the possibility of formally recognizing and affiliating with other professional groups of teachers which embody similar standards. Reciprocal membership is granted to members of any recognized society who change residence from the jurisdiction of one society to another. The benefits of this amendment are twofold. First, it paves the way for the evolution of national or regional societies made up mainly of STAT teachers like the ones already in Canada, Australia and Switzerland. As these societies ultimately become autonomous, it provides for the approval of new trainings, the certification of new teachers, and the day-to-day business of a society to be carried out locally where the issues occur and the facts reside instead of at a distant and already overworked central location.

Secondly, it also provides for the first time the possibility of a formal professional affiliation and acceptance between STAT-trained teachers and non-STAT teachers. This is where we come to the importance of the other recent event — the formation of the new *North American Society of Teachers of the Alexander Technique* (NASTAT)[§§]. The Society began as the germ of an idea in May this year and had its first By-laws announcement meeting on July 27th and a formal announcement at the Congress on August 10th. While the Society still has many details in the By-laws to work out there is agreement in principle on the main standards of training and membership. It will consist of an Eastern USA and a Western USA region (and hopefully a Canadian region if the Canadian teachers wish to join).

Evolving out of the same impetus as the motion into STAT, the new society also has a two-fold purpose. One is to link together teachers trained by the two existing professional bodies in North America, namely those who trained in the various schools under the banner of the *American Center for the Alexander Technique* (ACAT) and those who trained in STAT schools, in such a way as to provide consistent standards of training in line with those in the rest of the world.

The other main purpose is to provide a way for those teachers who are neither affiliated to any existing society nor adhere to any traditional standard of training to become

§§ NASTAT has recently changed its name to AmSAT (The American Society of the Alexander Technique). I was the chairman of the committee which formed NASTAT and the author of most of its original bylaws, as well as the mechanism which later became known as the Affiliated Societies whereby various national Alexander Technique professional organizations share a common standard and recognize each others' trainings and teacher certifications.

part of the larger profession. It has always been relatively accessible for STAT teachers to work with non-STAT teachers or to take workshops given by other teachers (for instance, those given by Marjorie Barstow) and many have done so. It has been similarly possible for ACAT teachers and many have done so. But because of the lack of professional acceptance it has been difficult for the unaffiliated teachers to come and learn in more traditional modes of the STAT and ACAT schools and few have done so.

What we have set up in NASTAT is a means for these teachers to spend some time in training courses so that they can add the repertoire of skills and the foundation of use taught there to what they already have. In the process they will get to know us and our work and we will get to know them and their work (and hopefully learn something ourselves). This is what it will take to be able to unify the profession in North America. With that unification will come the linking to the teachers in the rest of the world.

Who knows how all this will work out, but at least there is a demonstrated interest in such a global profession and mechanisms for supporting that interest are being put into place. It just remains for us to use it.

Recent Developments
Reprinted from *The Alexander Review, Vol.2 No.1, Jan. 1987*

I undertook this column with a view to using the space to look at the practice and development of the Alexander Technique from the broad perspective — with an overview. From a vantage point of travelling a lot and keeping in touch with teachers and trainings in many different countries, I had initially envisaged this leading into ruminations upon developments in teaching, new territories of application being explored, comparison of stylistic approaches, etc.

There is much to say upon these subjects and others which is of great importance to us all, especially those teachers who are not able to get around to work with their colleagues who teach in other modalities or other places. The Technique, whichever way one has learned it, presents endlessly ramifying seams of gold to the ardent miner, but it doesn't take much exploration in conjunction with other teachers to discover many other sorts of precious material which we didn't even know existed. Future columns will dig into all this richness.

However, as it has turned out, the last several columns have been pre-empted by momentous events, not in the practical teaching side of the Technique, but in the professional, political side. These events will come to affect us all and so it is important that everyone is kept as fully informed and as current as possible. Being in a place near the centre of many of these events, I hope I can get away with using these pages to describe them and their implications, since the Review gets to more teachers and trainees from more different backgrounds than any other media.

News travels quickly via the grapevine in cities where there are numbers of teachers, but much more slowly to those teachers out on their own spread thinly across the countryside. To communicate with teachers and keep them current STAT has its Newsletter, ACAT-West and AuSTAT (the Australian Society) are both to be congratulated on their publications, the Swiss Society, the Canadian Society and probably others also send out newsletters, but for the most part these go out only to their respective members and associate members.

It has now been a year since the inception of the Review, during which there have been growing pains. But thanks to all of you, it is the closest thing we have to a profession-wide and world-wide forum with a circulation approaching 600, made up of teachers from all over the world. So in the interests of getting the news out as soon as possible to as many as possible, I'll carry on in this issue with some of the more recent developments.

As mentioned in the previous column, both STAT and NASTAT (the newly formed American Society) have included in their by-laws formal means to open official recognition of other bodies of teachers (including each other). Ultimately this will lead to reciprocal membership between such recognized societies when members move to each other's territory. While the machinery has yet to be used between STAT and NASTAT since NASTAT is still in the final formative stages, it has already begun to influence the shape of things to come.

Other societies are beginning to use it as a model for their own by-laws. There currently exist six other societies of teachers in the world (as far as I know, correct me if I'm wrong): the Canadian Society (CanSTAT) and the West German Society (GLAT) have already expressed their intention to become fully autonomous and join a league of affiliated societies as soon as possible; in addition there is also a Swiss Society, a Danish Society, the Australian Society (AuSTAT) and an Israeli Society many of whom will take this path soon, if they have not already begun. ¶¶

At the second part of their annual general meeting in October, STAT voted, on a motion proposed by Walter Carrington, to define a specific territory for itself as regards reciprocal membership with members of a recognized society. Whereas, previously, the section on reciprocal membership had offered the possibility of STAT membership to anyone moving out of the territory of a recognized society to anywhere else not already covered by another recognized society, now reciprocal membership is open to members of a society recognized by STAT only if they should take up residence in the United Kingdom or Ireland.

This is a much more workable proposition for an already over-burdened STAT Council and dovetails with the larger moves by STAT toward devolving from being a world-wide body which is what engendered the whole idea of recognized and affiliated societies in the first place. Anyone now moving from the territory of the society in which they are a member to a country where there is currently no affiliated society would remain a member of that society. This will share among the various societies whose members move about in the world the difficulties of communication and representation which are now almost exclusively in the not-so-coveted custody of STAT. It will also spur on those teachers in such an area to band themselves together, the better to meet their own local needs, rather than leaving them isolated expatriates of a distant organization.

In December, the STAT Council ratified a paper drawn up by the Chairman, Margaret Farrar, and myself outlining the procedure of affiliation from STAT's point of view. This has been sent to the heads of all the other societies and to the directors of all STAT training courses. It institutes a six year transition period, beginning immediately, during which, in any country where a society wishes autonomous affiliation with STAT, the STAT members and STAT training courses there can keep a foot in both camps while the local society establishes itself. After this period the local society will be expected to look after

---

¶¶ There is a complete and current list (as of February 2012) of the Alexander professional bodies in the Appendix at the end of this book.

all its own affairs, including acceptance and certification of members, and regulation of present training courses as well as approval of new ones.

During this period a training course approved by STAT may choose to be approved by and have its graduates certified by both STAT and the local recognized society, or by either as it sees fit. The only restriction being that the director would have to choose which society would be called upon to regulate affairs in the case of a problem arising so that there is not a confusion of jurisdictions. After this period, the training would only be able to be approved, certified and regulated by the local society. So it is obviously to the advantage of any training to get involved in the local society so that in 1993 the transition is not abrupt.

A STAT member during this period will be still a full member with voting rights and the right to hold office (provided, of course, they pay their annual membership fee) as well as being able to become a full member of their local society. After this period, the member would still be able to be member of STAT, but as an Overseas Member at a reduced membership fee and without voting rights or the right to hold office. So again, it is to the advantage of any member to become involved in the local society as soon as possible in order to have a say in the evolution of that society.

It is hoped that this transition period will encourage the evolution of the societies other than NASTAT which are mostly made up of STAT members by giving them breathing space to begin a full range of operation in tandem with STAT. It will also serve the purpose of allowing the STAT teachers within the area of the local societies to see over a period of time that the standards of the societies, their certifications and membership are the same as or equivalent to STAT. In fact, the very idea of the league of affiliated societies is that they all adopt a common standard and so their certificates represent that common standard. The next few years should be very interesting…

Looking Outside Ourselves

Reprinted from *The Alexander Review, Vol 3 No 1, January 1988*

I recently attended and gave a talk at a conference in Northern Denmark sponsored by the Fölke University in Aalborg and organized by Chris Stevens who directs the STAT-approved International School for the Alexander Technique there. The purpose of the conference was to acquaint the Danish medical profession with the Alexander Technique — its practice, its scientific foundations and recent discoveries relevant to the Technique *.

The conference began with a showing of the film, *Posture and Pain*, a shorter, re-edited and much superior version of the Channel 4 film, *A Way of Being*. Chris then gave an overview of the background of Alexander and the Technique and of research into the Technique, including that of Dr. Barlow, Frank Pierce Jones and his own work. Kathleen Ballard, an Alexander teacher with a background in neurophysiology delved into the relationship of muscle spindles, the organization of the nervous system and our thought and direction. There was then a period of hands-on Alexander work to give the attending doctors an experience of the practice of the Technique.

After lunch, Michael Nielsen, from the University of Aarhus, detailed his research showing that the Alexander Technique is as effective in reducing high blood pressure as beta-blocking drugs. Dr. Finn Bojsen-Moller, professor of anatomy at the University of

Copenhagen, fascinated us with his work on the elastic properties of connective tissue and how it adapts under stress so as to send unreliable sensory messages to the nervous system. I finished by bringing up the issues I wish to elaborate below then illustrating them with some of the material which I've been getting into in the series, *In Our Own Image*, in this journal.

Conferences like this one are very important to us in several ways. Especially when they include participants from other disciplines than our own. They bring to us new discoveries and new models developed to represent these discoveries. This presents a dual challenge for teachers and students of the Alexander Technique. It is essential that we sit up and take notice of the world out there; that we get serious about finding out about and incorporating new research which is of relevance to our work. For if we don't, at the very least we shall be operating with old, probably outmoded, ideas and hence cannot take advantage of the insights and information more recently uncovered. At worst we will be operating with outright misconceptions and thus, inevitably, our work will be inherently flawed.

It's not comfortable nor easy to take in new ideas, to have to shake up and rethink part of the foundations of your skill, probably just when you were getting it all coherent and cohesive. But our work can only improve and we can only gain by remaining flexible and non-elitist in our approach.

In addition, it is important that we, as teachers (and future teachers) are able to understand our work in relation to the modern modality of science. For our work to continue to grow and become accepted we must be able to, for instance, stand up in front of a physiologist and describe what we do without appearing a fool. If we want people to know what we do, *we must know what we do*, in the sense of being able to adequately explain it. It is our job to ensure communication by being able to speak and listen in the common language of everyday speech as well as the common language of science, even while we introduce our own terms to invite people toward an experience of what we do. This, to me, is what it means to be a professional, colleague of other professionals, rather than an amateur with my head in the sand.

The secondary challenge for us is built upon the first. It is important to realize that all our conceptions are just models. The history of scientific revolutions is the story of new models which more fully describe the world and experience replacing old ones which were once widely accepted but now can be seen to be limited or just plain erroneous. But it is also the story of new models which open up whole new provinces, not necessarily replacing anything but augmenting and uncovering new territories of experience.

We have a method in the Technique in which we are trained to a high level of sophistication, particularly in using our hands. We know the value of this work in practice. We know what it opens up for people and how it gives them command of themselves in areas they didn't even know they had areas. If we are asking our students to go through such a radical change in their manner of use, we must provide them with a conceptual basis for that change which is, at one and the same time, in line with the principles of the Technique and in line with current knowledge.

That is to say, when we make the effort to take in discoveries around us and digest them in their own terms, we must then re-conceive and re-interpret them in *our* terms and hand them back to the world with the added value of having put them to the service of facilitating a whole and real person making a constructive change in their life at the present moment.

Most of us are not regular attendees at conferences and symposia, but at your local library, if you dig around in the multitude of scientific journals you will find a myriad of extremely interesting and relevant facts. You will also find many of these facts embedded in experiments with the most amazingly myopic goals and couched in the most numbing jargon. This is, of course, the price you must pay to get at the jewels, but it is also an indication of the nature of the model from which these facts emerge.

To paraphrase what Kathleen Ballard said in her talk: the design of many of these experiments is such as to be either totally removed from actuality (for instance, electrical stimulation of muscles in ways which never occur in living creatures), or at least missing the better part of actuality (experimenting on a tiny part in such a way as to exclude the effects which would normally occur because of interaction with other parts). All this serves to hide the direct and personal significance of their results from the experimenters in favour of a contribution to some more general and abstract model.

We, however, have a model of functioning and the initiation of functioning which is exceedingly different from the common cultural modality. When we take information from someone else's model and plug it into our own, the models do not necessarily need to be at odds with each other, though we often end up framing it that way. One does not need to be wrong for the other to be right. Neither may be wrong and both can be right. In fact, right and wrong don't really enter into it. The models can be seen simply as different points of view of the same reality, different ways of looking at the same thing, each necessary to the other and each appropriate to its own realm.

Perhaps the best terms to differentiate between these models are "subjective" and "objective". From the point of view of modern science, these terms have come to acquire unequal and opposite connotations. The "objective" in science is regarded as a positive virtue, connoting the real and true, the provable, that which is beyond personal bias. The "subjective" is suspect, that which is subject [sic] to illusion, projection and self-deception.

But they need not have these connotations. Subjective and objective merely cover different territory. The objective operates in the realm of how things work seen from the outside. It gives us knowledge appropriate to acting outwardly, making changes in the world — it gives us the mechanisms we can manipulate. The subjective is the realm of oneself, the conscious individual, with intention and volition operating oneself from the inside — it reveals to us the self we can use (to manipulate the mechanisms — if we choose). Obviously we need both to make up a whole.

In other words (and it may sound almost too obvious to say it this way), there is a lot going on outside the Technique of value to us, but *we* also have a lot to offer. What we have above and beyond our ability to help individuals improve their use is an approach, a way of seeing things, a means. It is a means which puts the meaning into things. That is to say, it gives us a value system to judge the significance and value of whatever we come across by bringing it into relation with ourselves and using our own psycho-physical system as a register.

What else is it for, all the sensitivity and awareness we can open up? Not just for perceiving ourselves and what's around us, but to go beyond toward actually apprehending what we want and what is good for us. This is what we have to offer — a means which

104

people can use to open up their systems to become a scales, a balance, a register uncovering the personal relevance, the value for them, of whatever they take in. More than this, in fact, since we are all much more similar to each other than we are different, it is a means for developing a register for the human nature we have in common.

The important thing here is that in our work we already have the practical method and the experience. We can benefit from the research of others by gaining better ways to understand and convey that practice. We also, through our experience, can interpret this research and better show the practical significance of it — a significance which is not always so apparent to those who discover the facts because they are usually not able to put them into practice in their own being.

Chapter Six

Experience & Experiments
in Alexander World

The following was first published by Direction Journal in
The Congress Papers. It is a description of a masterclass given
by David Gorman at the 3rd International Alexander Congress,
which was held in Engleberg, Switzerland, August 1991.

As everyone who was at the recent Congress in Engleberg could see for themselves, we are starting to grow up as a profession. Not only in terms of numbers, but also in the confidence and willingness to look around us at our colleagues and to marvel at the range of talents and interpretations of this work we do. Except for a few sourpusses who held themselves aloof in the mistaken idea that their technique was the real thing and needed protecting from those who were either charlatans or fools, we all got stretched and inspired, appreciated for what we can do and able to see what we might do.

This seems to me, having attending all three congresses, the first one which really included and was representative of the entire community — by the entire community I mean all those who are sincerely exploring the Alexander work and teaching others (and of course, therefore have something to offer us all of their knowledge and their discoveries). It is precisely those who are farthest from us in their background and their way of approach from whom we stand to learn the most. I was very pleased to see so many people so excited by so many different understandings of the work. We still have a long way to go, both in opening ourselves up to the people around us in our own profession and in developing the potential of the work, but we're off to a good and promising start that feels more solid than ever before.

It is a particularly good time for us all to be coming to this consolidation of the meaning of the work through sharing and openness because the profession has reached the "age" when those who worked directly with Alexander are retiring or passing away and the mantle is passing to another generation of teachers who do not have that direct remembrance of where the work originated. We have only each other now and what we have learned from our teachers and discovered for ourselves, so it behooves us to begin to make the best of each other so that we don't find ourselves slipping down the path to discord and dissipation of the force of the work through conflict in a way that has afflicted so many other nascent professions. Anyway, that's enough of a plug for tolerance and openness…

The congress was such a fullness of possibilities that few of us were able to attend more than a fraction of the groups and workshops we would have liked. Perhaps as we read through these Congress papers we'll be able to glimpse at least something of what we missed and whet our appetites for Australia in '94. I had initially decided against accepting the invitation to give one of the second generation classes because I've become less satisfied with the value of workshops where I show how I teach by working with others and demonstrating how I do it. It has become much more interesting for me to help others find their own way of teaching and/or what is stopping them from developing their unique and individual expression. I couldn't quite imagine how to approach this in the time and format available.

Then I remembered what one of the students on my training course had said about a lesson she'd had with Peggy Williams. The student had asked Peggy what she did to get ready to teach as she was about to work with a pupil. She said that Peggy had replied something to the effect that "why should I do anything to teach, I *am* a teacher!" It made me realize, of course! Why should we do things to get ready to teach if we've already integrated the work into our daily lives. We're as ready as we're ever going to be. If we haven't integrated the work into our daily lives then we're not suddenly going to get any better by preparing for a few seconds — we're only going to get a little more prepared and less our normal selves. If we feel we have to get ready before coming into contact with a pupil maybe we should work a little more on integrating our "good use" into our daily lives so that we actually are living what we suggest to our pupils.

So I decided to use the second generation class as a way to experiment with this issue with the various teachers and trainees who attended. We kept it simple — just splitting up into small groups, each person taking turns to come up to work on another as they "normally" did, the others observing to see if they stopped and got ready, or "directed", or "released" or anything special that happened just before coming to contact and proceeding with the lesson. It was surprising for a lot of people how much they put in between them and the pupil in terms of preparation.

The second part of the experiment was then to leave out all that intermediate "Alexander stuff" just to see what would happen. Here it was interesting how difficult people found it not to do their usual "teaching" stuff. They felt as if they were no longer doing the Technique, or that they couldn't possibly teach, or that they would be no good without their extra "Alexander armour". It revealed a lot.

But the most fascinating (and powerful) thing was what actually happened when the "teachers" didn't do all their "teacherness". The "pupil" being worked with and the observing group all could see and feel the change. In their own way each person felt that the teacher was "more with them" as opposed to behind their teacherness. They felt more "allowed", more "space and warmth". It was as if the teacher by being more themselves allowed the pupil to be more themselves, which felt good and was appreciated. The pupils found themselves less concerned with what was supposed to happen and less anxious of what was expected of them. They were more present with the rest of the group and less drawn in to some inner physical feeling process. For most participants this was all quite unexpected and interesting. Once the teachers got over their difficulty in letting go of the perceived necessity to "direct", etc. they also felt more at ease and had more enjoyment in what they were doing. They didn't have to *do* the teaching, they could *be* the teaching. In other words, they could *be teachers*.

I think for many it was a surprise to realize that, indeed, they already were highly-trained, sensitive beings embodying a lot more of the work than they had thought. That their preparation and doing actually kept them from their own integration. It distanced them from the pupil and from the human responsiveness between the two of them. I'm only sorry the time was so short (and that I had to miss the other classes to do my own).

I think I'll go as a "civilian" next time…

Chapter Seven

The Rounder We Go, The Stucker We Get

A Brief Introduction

I have updated this article for the new edition of this book. I originally wrote it back in about 1996 from a talk given in 1993. I could have left it as is, as a historical record of how I was seeing things at the time I wrote it. It was intended then to be a map of the nature of circular habits and how they seduce us and I wrote it in the hope that it would help others see through their own versions of this habit.

I think what this article has to say is still very important. However, my understanding of the nature of circular habits has evolved a lot in the twenty-something years since then. Rather than let the old map pretend it still accurately describes the whole territory, I figure it is more appropriate to touch it up a bit, correcting a few parts, making it more detailed in places, and filling in some of the blanks, all so that it can do an even better job of guiding others.

Nevertheless, it is still by and large the same article. The changes I have made are not huge. Of course, if I was to rewrite it from scratch today it would be a very different piece. What this means is that some parts of the article have not been changed so much as they have been restructured and rearranged. Some other parts I have rewritten, and a few bits have been added as entirely new text. I have also modified some of the illustrations as part of the update.

Let me be clear. The main point of this article is still as true to me now as it was then — that we do not really see how we entangle ourselves deeper and deeper into the circle each time we think we are doing something to get out of it. It is still just as important to uncover how we unwittingly take the next step in reinforcing the circle, and then stop doing it. What is new to this present version is that I have added a few glimpses of how we can also change what causes these habits in the first place.

Read on…

The Rounder We Go, The Stucker We Get

The Nature Of Circular Habits
and Hints on Escaping Them

Originally titled *"The Rounder We Go, The Samer We Get"*
From a talk given in London, U.K. in May 1993 at *The Centre for Training*
(an Alexander Technique Teacher Training program run by the author)

This is about the nature of habit — unconstructive habits, particularly that kind of habit known as a *vicious circle*. That is, a reactive cycle, each step of which brings up something that forces me to react by taking the next step, which in turn forces me to take a further next step, and so on, inevitably binding me into repeating the cycle, thereby reinforcing it and making it into a habit.

No one wants to be caught in vicious circles and certainly no one wittingly sets out to trap themselves in such destructive habits. Nonetheless, we do find ourselves caught in them and unless we understand how they work we cannot step outside the cycle that reinforces our existing habits nor can we become free from creating new ones. If you (or someone you know) are stuck in a habitual and chronic "problem" which you cannot eradicate in spite of your best efforts and the most you have managed is to find better and better ways of dealing with it, then you have your own concrete experience to refer to as we go through this.

The key to understanding how we unwittingly create and sustain these kind of habituated cycles is to see how they are built at every step upon an integrated series of delusions which both drive us and seduce us into taking the next step. What we think we are doing to *solve the problem* seems to make sense within the context of the habit, but from a larger perspective is, in fact, the very way we *create the problem* and sustain it. These habits are brilliantly constructed in the way they trick us into repeating them in the face of our resolution to change. Even more cleverly, the way out is hidden in the very last place we would ever think of looking for it. Not only that, but in understanding the structure of circular habits, we can begin to make sense of why the habit of "trying to solve the problem", of trying to be "right", or "perfect", or "ideal", or even to become "different" has such a powerful hold on us, for we are deeply caught in trying to be in control.

Let us explore how all this works, using, for ease of understanding, the very common example of chronic, uncomfortable "tension". Tension is a symptom that shows up in a mostly physiological way (that is, in body feelings as opposed to in an emotional or psychological way), but it will become apparent that these kinds of circular habits manifest in every aspect and territory of our lives. I will start with the nature and mechanics of the habit and how we get caught in it, then show the unexpected and hidden doorway out of it and finish up with some of the implications of bringing about fundamental change.

1 THE NATURE OF THE BEAST

I go about my life, doing this and that and not thinking much about *how* I go about what I am doing — I just do it. Then, in one particular present moment in the middle

In this present moment the symptom (pain or tension) feels WRONG and I react to the feeling by immediately wanting to do (or undo) something to change it

of whatever I am in the middle of, I am brought to consciousness by a "symptom" — some feeling of discomfort, of pain, tension or the like. This symptom appears, grabs my attention and I naturally feel it as something wrong, something I do not like, a "bad thing". It is also natural for me to feel that this symptom *is* the problem and, equally obviously, to want to do something about it to make it go away; to make everything better, preferably as quickly as possible. Success, of course, would be to get rid of the symptom and go back to what I was doing, *minus the problem*.

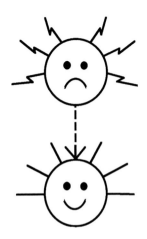

In the moment of awareness the symptom (pain or tension) feels WRONG and I am forced to react by...

...immediately doing something to release it, change it, get rid of it, so that...

...in the next moment I will feel RIGHT and OK again and I can get back to my life...

So, I go about some process or some act to change the moment from an "*it's-not-OK-I'm-feeling-bad*" kind of moment to a "*now-it's-OK-I-feel-better*" kind of moment.

Let us take the example of me drawing a picture. After some indefinite period of working away with great concentration, I am brought to awareness in that present moment by a feeling of soreness and tension around my shoulders which (of course) feels bad to me so I want to do something to relieve the symptoms. Maybe I tighten up my shoulders, pull and scrunch and mulch them about hoping to break up the tension and ease the pain. I might try to release the tension or relax my shoulders. I might massage the sore spots or get someone else to do it for me, or whatever...

It is important to recognize that in terms of the habit it does not really matter what particular process I use (whether I do or undo something), the point is that the unpleasant feelings of the symptom force me to react by trying to get out of this present moment and into the next moment when (hopefully) everything will be better. When I succeed in making the *wrong feeling* (the tension and pain) into the *right feeling* (no tension or pain), the *not-OK* moment into the *OK* moment, I can merrily get back to my life without having to pay attention to all this stuff any more because at this moment it seems to me that there is no longer a problem.

All of this appears to make perfect sense and would be of little consequence if this was the first and only time these symptoms occurred. I probably would not even think twice about it. However, that is *not* what happens to most of us. What *does* happen is that after making everything better I just launch back into my life again, but it does not take long — a few hours, a few days or weeks — before the symptoms are back. I, of course, immediately react to the *wrong feeling* of the symptom by doing what I have always done before which to is to *do something* to get rid of it. Which works most of the time to get

113

me back to where I feel OK again. And all would be fine if I stayed there, but, sure enough, the symptom is soon back. Not only that, but over time it's getting worse and what I used to do to change things for the better does not work as well any more.

For a long time I never question the possibility that there may be something funny about my whole approach, I just try to find another "better" way to make everything OK. So I try this or that method and it works for a while or I try this or that exercise and maybe it "works" and maybe it does not. But I keep on trying, and the symptom keeps on recurring until I get so familiar with it that I begin to regard it as "my problem". I find myself starting to think of what's happening as an "it". Me and *my problem*. There *it* is again. *It* hurts me. I *have* tension. I *have* a back problem. I would be fine if I didn't have this #%@#!! problem...

I come to awareness of the symptom (pain and tension) which feels WRONG and/or BAD, so...

...I do something to change things so that in the next moment...

...I'll feel GOOD and everything will be all RIGHT again...

...but it doesn't seem to last long because soon...

...there is the symptom back again, once more feeling BAD and/or WRONG...

...so maybe I try something different to change things so that the next moment...

...I will feel RIGHT and OK again and can get back to my life...

In fact, what is really going on is that "it" *has* me. There is an inevitable sequence of events that becomes established: feel wrong → react in order to → feel right so I can → go back to "normal" life → until I feel wrong again → and react again to → feel right → to go back to "normal" life... and so on, over and over, until this sequence itself becomes "normal life". Gradually the symptoms get worse and more persistent and I am forced to try different ways to "solve" the problem, but to no avail. I am well and truly stuck and no matter what I do I cannot seem to change anything more than temporarily. This, of course, is the situation many people find themselves in.

By this time it has begun to dawn on me that this sequence of events is not a linear sequence but a circular one — I had the tension, I did something to get rid of it and everything was OK, then I find myself back here in the symptoms *once again*. I am caught in a vicious circle like a noose gradually tightening around me the more I try to escape it.

At this point I might just resign myself to "having" this problem. "It is my tough luck, I have a weak back" or "we are not evolutionarily designed to do this kind of activity." However, I am still stuck with the recurring symptoms so I am forced to go on searching for better and better ways to get rid of them. Maybe I should try a different method or a new treatment, or maybe do positive thinking and affirmations.

Is this situation familiar to you? Are you intimately acquainted with the little beast? Of course, all the above would be wonderful if you had actually managed to rid yourself

114

once and for all of the problem and its symptoms by means of some treatment, some therapy, some medication, some meditation, a change of chair, a different diet, whatever…

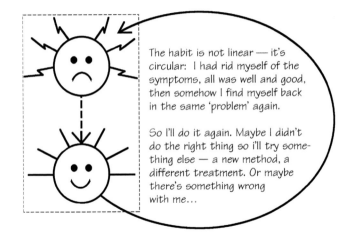

The habit is not linear — it's circular: I had rid myself of the symptoms, all was well and good, then somehow I find myself back in the same 'problem' again.

So I'll do it again. Maybe I didn't do the right thing so i'll try something else — a new method, a different treatment. Or maybe there's something wrong with me…

If you have, well, you are one of the lucky ones because for 15 years I have been working with people caught in just these kinds of habits and for most of them nothing they have done and no method they have tried has actually freed them so far. They may be getting better at "dealing" with their symptoms and habitual patterns in the sense of having better tools to make the change from the moment of the symptoms to the moments of relative freedom, but always they find themselves back in some similar sort of problem that needs dealing with. ***

2 STEPPING OUT OF THE CIRCLE

A number of years ago, as the nature of these circular habits was becoming more clear to me, both from my own experience and that of my pupils, it began to dawn on me that something was missing. Surely it must be possible to truly rid oneself of a habit rather than just get better at managing it, at getting better coping strategies? Alexander certainly stated that he had become free of the symptoms which had plagued him from childhood. So I determined to find that elusive way out of the circle. It struck me (in that obvious way that seems so obvious once it is obvious) that I get caught in circular habits because I keep going around the circle!

At each moment I somehow manage to take the next step even though I do not want to end up where it leads me. The odd thing is I do not feel like I am are taking a step deeper around the circle each time. Rather, I am desperately trying to take steps *out of the circle*! I figured that somewhere in there I must be missing something; somewhere in there must be the clue to the trap. Somehow I am taking the next step in the habit while being seduced into thinking that I am taking a step out of it. But how could this happen? How could I be so fooled?

I went back and re-examined each step from this new point of view and, sure enough, there it was. My habitual way of perceiving things had completely hidden it from me! Go back to the diagram of the circle (reproduced below). There is one time in that circle that I am "naturally" brought to present-moment consciousness — the moment of the symptom. In fact, *I am brought to consciousness by the symptom*, by the feeling of tension or discomfort.

*** Or as another Alexander teacher, Barbara Conable, said: "Releasing tensions is like swatting flies, there's always more."

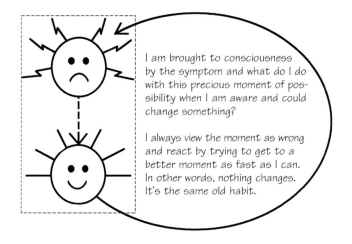

I am brought to consciousness by the symptom and what do I do with this precious moment of possibility when I am aware and could change something?

I always view the moment as wrong and react by trying to get to a better moment as fast as I can. In other words, nothing changes. It's the same old habit.

And what do I do each time?

I do exactly the same thing I have always done before — I try to get out of that moment which I regard as *wrong* and *bad* as fast as I can into the next moment when things will be *OK* and *good* again. Every single time! No matter what else I change, that part stays constant and I never question it.

That one precious moment when I am conscious enough to make a really different choice, I do not change a thing!

At that moment I believe I am taking a step toward the "solution". But there it is, plain as day: those steps from the "problem" to the "solution" are the same preconceived habitual steps I always try to take and are just as much part of that circle as any other part. For all that I think of it as the "solution", it is, and has always been, an inextricable part of the habit. It is the warm and lovely seduction which I cannot resist. At that moment, the *not-so-nice* feelings are driving me from behind and the *oh-so-nice* future is beckoning me from in front and I simply cannot conceive of any other possibility than getting away from the one and to the other (and I would not be attracted to any other possibility anyway). It is all so beautifully constructed and so cleverly hidden from me!

The habit is laughing (if I personify it for a moment), "What a sucker, he fell for it again! No matter how many times he has been through it before, up comes that wrong feeling and *WHAM!* before he knows it, he reacts in exactly the same way he always does and then he wonders why he is caught in a repeating cycle. Har-dee-har-har!".

The whole habit is set up so that, whatever I have been up to previously, the moment of coming to consciousness of the symptom is presented to me as NOT GOOD, as WRONG. I *feel* it that way. And I fall for it every time and react as if it *actually is* wrong and therefore immediately try to change things. I am totally and utterly convinced that those feelings ARE wrong and I MUST do something about them, thereby inevitably taking the next step in the cycle. In fact, without realizing what I am doing, I positively and desperately *want* to take that step back into the habit each time. I am *begging* to take the next step. I am *searching for better ways* to take that step! What I also do not realize is that, in effect, I am being *forced* to take that next step because no other alternative is imaginable. That is to say, I am forced to *react* to my feelings. This is what reaction is — *because of this… therefore that*. No other possibilities, no choice.

Notice how integratedly the habit works and how much "*of a piece*" it is. Reacting to the moment of the symptom and quickly stepping out toward my ideal "end" of feeling OK puts an "end" to my presentness and choice, for now I automatically begin to go back to being involved in my "normal" life. Which, of course, means that I now go back into that narrowed state where I am only conscious of *what* I am doing, not *how* I am doing it.

That narrowed "unconsciousness" is as built into the cycle as everything else.

This is important. This thinking that I have finished the job (for now) by getting to the nice place (for now) is what keeps me from using the wake-up call of the symptom to look deeper into why i have the symptoms in the first place. More on this later…

Instead, every time the symptom pops up, I think that *it* is the problem and I immediately try to get rid of it, to fix it all up in the preconceived way I think it should be fixed. All I am doing is developing more sophisticated coping strategies but I have not liberated myself from the problem. (And I must stress here again, that it does not matter what my idea is of what should be happening, it is always the same *fix-it-up* reaction in general and is always as much part of the habit as is the so-called "problem".) Each part of the cycle validates and leads to the others and they all fit together like well-oiled gears in a machine — an assembly line for problems.

Because I have been in that narrowed state through much of the time as the habit cycles, I am not able to see how the whole pattern works. At each step I can only see as far as the next step. The consequences of what I am doing are always just over the horizon and therefore invisible to me. As Isaac Dineson (Karin Blixen) wrote in *Out of Africa*, "The earth was made round so that we would not see too far down the road."

When I recognized that I had been deluded into buying into the whole scenario as presented to me by the habit, I did not quite know where to go. Everything seemed to be "lies". To paraphrase Alexander, "If anyone was in a rut, it was I". I had gone far enough to know that trying to make things OK was what drove the habit and that searching into those "unconscious" moments was like examining all the wonderfully involving details of the walls of the rut — fascinating, but I am still in the rut. The real trick would be to simply step out of the rut. But how? That is what I had been trying to do all the time and it did not seem to work.

I had no choice but to go back to examine the whole thing again in the light of my knowledge that nothing was as it seems. I kept being drawn back to the moment of the symptom. After all, it is the occurrence of the symptom that first lets us know we have a "problem". It is only because of the symptom that we are looking for the *now-I'm-OK-again-all-is-good* "solution". We never seem to question its reality because it is an actual sensory experience.

A moment's thought, however, will show that I feel the way I do at this moment because of what has been going on during all the accumulated moments before. And that what has been going on in all those moments before is that I have been cycling and recycling this habit. And that here I am about to give it another go round. Here I am about to crank it around one more notch. What do I expect to happen when I do the same thing I have always done? Something different this time? How deluded can you get?†††

It was then that I realized with a shock that the answer was right in front of me. The way out of the circle was simply to meet the moment of the symptom which I habitually feel as wrong and not take the next step of reacting to it. To accept that what is going

---

††† "The true relation of "cause and effect" on a general basis in connexion with the working of these mechanisms will not be given due consideration and, as we shall see, the majority of effects (symptoms of some "cause" or "causes") that they chance to recognize will not be treated by them as such, but as "causes," and dealt with in accordance with the "end-gaining" principle." (F. M. Alexander, *Constructive Conscious Control of the Individual*, page 186, Gollancz 1987)

117

The way out of the circle was simply to meet the full force of the moment of the symptom which I feel as wrong...

...and NOT to take the usual next step of reacting to it by trying to get somewhere that seems better...

on at this moment *is actually going on* — whether I like it or not. In other words, to freely and willingly live in that moment no matter how I feel about it, *simply because it is there.* I saw that this moment is just as valid a moment of reality as any other. What makes me think any particular moment is wrong and should be fixed up? Only my feelings. Very strong feelings, admittedly, but nevertheless feelings which have themselves been created by the cycling of the habit and so are automatically suspect. And who am I, the person who is so thoroughly caught in this habit, to know what is the "right" thing to do? Such unwitting arrogance... And so trapped and manipulated in the slavery of feelings that only the moment before this simple choice of acceptance had been totally unthinkable.

As this startling realization came to me and at the very moment of recognizing this acceptance as a possible option, up rose the total and utter conviction that it would be pointless to try it because to decide to stay in that moment would be to stay stuck in the symptoms (probably forever). No way! But, if I could just get to that next moment when everything is OK again, I would be happy to live there forever, but not here in this mess, thank you very much.

This was the habit, of course, like a little devil on my shoulder, trying to convince me that my plan of acceptance could not possibly work — *that it was unacceptable.* And I almost bought it until I remembered the profound shock I had experienced when the insight had first come to me. There was something so unexpected about it that I knew it was truly new and I also knew that I had never even dreamed of carrying this out before, so how could I be so sure what would happen. So I did.

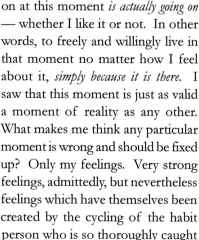

What did actually happen when I *inhibited* my monstrously powerful habitual urge to react to the feelings of the "symptom" was very, very different than I expected — after a moment of intense awareness of narrowness and restriction (during which I had to choose again and again not to react), an expansion filled me up and the strain and the tension disappeared! I was in a state of wholeness and oneness with myself and very sharply and vibrantly *present* in the world around me.

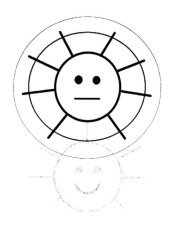

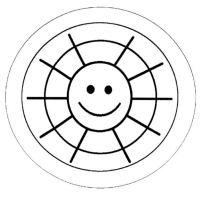

I felt good — in fact, better than good — for along with the expansion came a rising delight and intense aliveness. All by doing absolutely nothing but meeting that moment I had always taken to be wrong and refusing to react to it the way I always had before. And lo and behold, it seemed the moment was not wrong after all — it was wonderful!

Which makes sense when you think about it... The moment I truly accept what is going on just for its own sake, is the moment I truly give up to it. I am no longer struggling and straining to get to a "better" moment. There is no longer a split in me with one part of me that feels another part of me as wrong and therefore not OK to be in the moment. There is no longer that one part of me trying to change that other part to what I think it should be (as if that were possible). Thus, I am no longer in a fight with myself at that moment. And, as we all tend to forget, *you can never win a fight with yourself* — *one of you is going to lose and it is always going to be you!* And it's going to be you simply because you engaged in the fight.

To make the *choice* to be in the actual moment I am in is also the moment that I give up imposing my deluded version of what *should* be happening and leave myself open for something new. And that is exactly what happens — something new. Ironically, what happens when I do not try to reach my wonderful goal is that I actually do end up with what I wanted. All the elements I was after are there — wholeness, freedom, openness, alert presentness, no strain or tension — but they are all there in a very different way than I had imagined or experienced before and, of course, I have arrived in this new "place" by a very different path, or rather, by no path at all.

One experience, of course, does not prove anything, but after many times in many varied situations, when I had managed to meet the force of whatever "symptoms" or feelings came up without reacting by trying to make things right and instead willingly and freely chose to live just in the moment, no matter what it was; and after roughly the same sort of experiences had resulted each time — I was beginning to believe it. Now, years later, I have helped hundreds of others learn how to make this choice too and their experiences are always similar — at the moment of truly giving up the *reaction* and *end-gaining*, a wonderful peaceful easing quickly spreads and expands throughout them and they open up as whole, breathing, supported beings to the world around them, present and ready to respond, all over, all at once.

In my own experience and in that of working with pupils, the biggest challenge for everyone is to meet that moment of feeling and not react by trying to change things. In fact, this is the biggest change it is possible to make — no change at all — since habitually we always meet that moment and try to change things. Though it sounds like a very simple choice to make, if you have not done it you cannot imagine just what a huge challenge it is to actually carry out in the face of the immediate and overwhelmingly "real" feelings, thoughts and emotions of that existential moment. It takes every bit of courage and clarity to stick with this *means whereby* when your entire experience is screaming at you, "WRONG! WRONG! GET THE HELL OUT OF THERE AS FAST AS POSSIBLE!" The fact that such all-encompassing changes take place almost immediately when we do manage to inhibit reacting shows how big a change it is to willingly accept what we feel is wrong and to freely live in the given "here and now".

3 WHAT IS NEW IS NOT WHAT IS OLD

These experiences showed me that it is indeed possible to step out of the vicious circle, or at least that main part of the vicious circle which as long as it is there prevents you from progressing any further.

To step out of the circle *does not mean* finding a better thing to do; it *does not mean* changing the symptoms; it *does not mean* end-gaining for the right solution; and it *does not mean* that anything is wrong right in that moment. All is as it should be simply because it is as it is. To step out of the habit *does mean* actually accepting the moment as it is (which means having faith that reality is not broken and does not need fixing); it also means giving up my fixed ideas of what *should* be happening (which means discovering what is *really* going on in the present); and it means allowing myself to experience the feelings I really have instead of trying to get to the ones I want to have   which means realizing in actual experience that if I do not react, *there is nothing wrong with me!*

What then were the symptoms all about and why had they come to feel so wrong? Was the negative feeling just my habitual perception of the moment or was something truly unconstructive happening but I was just misinterpreting it. This seemed the next big question to tackle — mostly because it was there and would not go away — a situation that I was starting to realize meant that something important was going on.

I began to recognize how much I was learning by being able to see with fresh eyes the simple and simultaneous *facts* of what is, rather than the *acts* of what I could do about it. So I went back to my experience to look again at the facts. Go back to the diagram of the circular habit (*right*).

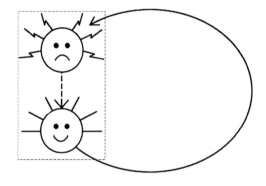

In this cycle I spend the majority of my time in a state of narrowed awareness such that I am not conscious of how I do what I do, I just do it until I become aware that I feel very tense and tight which, of course, is a physiological state of muscular constriction and contraction. Indeed, if I am a complete unity then how can my attentional contraction be in any way separable from my muscular contraction? I become a completely integrated whole-system narrowedness.‡‡‡

If, every time I am awakened from these substantial periods of being so narrowed, I discover that I have become cramped and tight then how can I interpret it otherwise? I

---

‡‡‡   By the way, narrowing into activity with such effort as to end up squeezing the life out of ourselves is not the only pattern, though it is one of the most common. There are many others, often with overlapping elements. For instance, some people find themselves split from themselves, usually their "mind" (the perceiver) from their "body" (the perceived). The perceiver doesn't like or can't accept the sensory and emotional feelings they get from the perceived "body", and backs away from identifying with and being themselves. This reaction, of course, is an emotional one, most often physically perceived as negative, which increases the person's unwillingness to accept and identify with such feelings thereby reinforcing the split. What the person does not realize is that what they are feeling is not their body, it is what it feels like to be so split. You cannot feel your "body". You do not "have" a body. You "are" your body. You are yourself all the way from bottom to top and out into the world. What you are feeling is what it feels like to be doing what you are doing, in this case what it feels like to be that split and that reactive; that much making yourself wrong and that much trying to manipulate yourself. And when this is a repetitive, consistent habit, the feelings will be consistent and repeated, therefore it is "natural" that you will become regularized to them and come to think of them as "you", as your "body". This is to say, in spite of whatever differences there may be in the content and details of the habit, its circular nature remains the same.

narrow (myself, my attention), I get narrowed (physically, my coordination). What I am feeling is not muscles contracting, not some bad sitting, nor any of a million little details. All of these details are undoubtedly happening, but they are not causes, they are effects. What I am feeling is what it feels like to have been so narrowed for so long.

Conversely, experience also shows that when I allow myself to exist in any moment as it is without reaction — in other words, to open more fully to the experience and events of the present no matter whether I like it or not — these tensions and contractions disappear and I become free and whole. How can I interpret that except that as I open, I open — in every way? More than that, I become more open and expanded all the way into that fresh alert awareness of the world around me than I was at that first moment when I made my choice to accept the narrowed "symptoms".

The implications flooded in on me. If the only thing that I changed was to NOT allow myself to react *as if the feelings were wrong* and consequently the symptoms disappear, how can I interpret that except that I am OK and always have been OK — I just did not know it before. I was deluded into thinking that something was wrong *in the present* and needed to be changed so that all would be OK *in the future!* It is important to recognize that the feeling of wrongness is not the delusion. There really is something "unconstructive" going on, but it is not the feelings of tension or pain. It is my narrowed, reactive, end-gaining state. There is NOTHING whatsoever wrong with the feelings. They are very real and valid feelings giving me very important information, namely, that it is a very constricting and painful thing to narrow off for so long into my drawing.

This soon led me to further insights. I had been working on a set of drawings and had a deadline. I had to finish several more pages that night to be ready for a workshop the next morning. So I was rushing, trying to get finished as fast as possible. However, I had not been drawing more than about a half-hour before I began to realize that I had an annoying sense of strain and tightness in my hand. "Oh, no, my hand is cramping up. That's all I need when I have such a lot to do."

My old habit years ago would have been to react to the strained sore moment by trying to turn it into a more free, i.e. not-sore, moment. I'd shake my hand, or rub it, or try to hold the pen differently. Anything at all to make it work better so I could get back to drawing and finish. More recently I'd have paused to free my neck or come to a more upright and poised sitting. But every one of those reactions would be part of that same old fix-it-up habit. The sole purpose of doing any of those things once I "woke up" in the bad place, would have been to get to a "good place". If I had succeeded, I'd have merrily rushed back into drawing with no more awareness of why my hand was cramped. Why should I? I'd just fixed it all up, hadn't I? Undoubtedly in a short while I'd feel the cramp again, fix it up again (hopefully), and go right back at it again. That's how drawing works, isn't it?

But that's not what happened this time since I now had more knowledge of the sneaky ways of the habit. Now, the symptom woke me up enough that I could remind myself to take a moment, to not automatically assume there was something wrong about the moment that needed fixing (by me). Instead, I chose to not react and to stay there just as I was for some further moments and not change anything. To just let myself be there, sore hand and all. In other words, to change what I was doing at the moment compared to the moment before, not to try to change what I was feeling the moment after.

The first thing that happened was that I noticed that I had time. My vision and aware-ness opened up and I took in more of the room. The moment seemed bigger and more spacious. That's when I noticed that I had started to breathe more and I felt myself settle more solidly onto my chair. At the same time I noticed an expansion around my shoulders and an easing in my arm. This was interesting. I had not "done anything" and yet my system was responding with all kinds of nice changes.

I'd had this happen enough times by then to understand why this lovely change had happened by itself. Of course I'd open up, and free up, and come back into the moment because I was no longer narrowed in and busy trying to change my sore hand into a good hand.

But it was more than that. The change also happened because I wasn't narrowed into the drawing and rushing any more. I was taking the time to be present and not out ahead into the next moment. This is when I recognized that by not getting busy trying to fix up my symptom, I'd also given myself the space and time to become aware of what I'd just been up for the last half-hour. I'd been narrowed into my drawing, rushing to get done. I'd been pushing things, actually trying to draw faster than I could easily draw.

The cramping and soreness was, in fact, the feeling of rushing. It was how my system became coordinated when I was trying to go faster. Or to put it slightly differently, my coordinating system was trying to carry out my intention to get the drawing done faster, but the only way it could make my hand and arm move that fast was to get the muscles working against each other to accelerate and decelerate the masses. The result of this was gripping and cramp in my hand that I had felt, and the tightening in my arm and shoulder which I had not felt. I had not noticed the interference with my breathing either until I stopped and it all began to release.

I realized with absolute clarity that all these effects had been caused by my rushing because I knew that that normally I could draw for hours when I was no hurry and have no strain at all.

I had not been tensing my shoulders or tightening my hand or stopping my breath-ing, though my shoulders had certainly become tight, my hand gripped and my breathing interfered with. What *I* had been doing was rushing, pure and simple. Trying to make my system go faster than it could do with ease and freedom. The tensing, tightening, and interference was simply the coordination of rushing.

And I proved it to myself because I now slowed down and just enjoyed the drawing and let it take the time it took. If I got it done, great, if not, I'd rather have a bit unfinished than hurt myself by rushing.

All this has big repercussions. That I come to wholeness when I stop reacting shows me that I *am already* intrinsically whole and integrated since that is what I am when I am just being me in the present — I am and always have been a "psycho-physical unity".

As I stop being divided by making one part of me wrong and trying to change it, I become whole. I do not *have* a mind. I do not *have* a body. I am me — the sum-total of my memory and my moment-to-moment experience — one indivisible whole. I was just too deluded to know it and hence was constantly engaged in precisely the kind of habitual reaction that was guaranteed to make me feel *as if* I was made up of a number of parts which needed some co-ordinating on my part in order to become whole in the future.

122

The experience of all-over release and ease is simply that it is easy to be myself as I am in the present since that is what I actually am when I am not straining to be what I am not. For what is on-going tension but an on-going conflict, a continual trying that cannot achieve its goal, like the rushing. It not only takes effort and energy to try to get into the next moment ahead of myself, it is downright impossible (though, habitually, this did not stop me from trying). How many of us have not yet learned or are quite unwilling to accept that we CANNOT be any different than we are, *and there is no more frustrated trying than trying to be.* I feel free *because* I have freed myself from the slavery of *having* to react to my feelings and the tyranny of my idea of what I *should* be.

By the same token, the expansion and openness that accompany my choice to allow myself into this moment shows me how narrowed I had been to what I was concentrating on in those moments before and how closed off I was to everything else that was happening. The profound sense of aliveness and presence in and of the space around me is the experience (beyond any theory) that I am not only *not separate in any way from myself, I am in no way separate from the universe around me*!

This brings me to an important point implicit in the above but perhaps worth making explicit. The circular habit is so all-pervasive that it even colours our deepest metaphysical and spiritual "precepts" (and what is a precept but a "pre-conception"). We all love the idea that we could escape from our split and conflicted prisons and become whole, integrated and part of the larger universe, but we still project that goal into the future. "I am not there yet, I am still in this mess, but I want to be there and with some exploration and learning maybe one day I will be free and where I want to be". On a larger scale we tend to see our human lives and cultures as caught in destructive patterns of behaviour and less than desirable moral and ethical interactions. We project a better state or a more perfect plane to aspire to and then devote our energies to trying to achieve them. Do you see the parallels? All these parallel tracks — these *let's-aim-to-get-better-for-the-next-moment/ life/ after-life* tracks — are heading in the same direction and the express train of our lives has wheels on all of them simultaneously. In other words, our fundamental mode of operating in this habit is "end-gaining" and every thought, feeling, emotion, and connection to the world around us will be interpreted and subsumed into this habitual pattern of reaction.

We conveniently "forget" what we all kind of, sort of, really know and that is that there is only the present moment. All past moments are already gone and no changes can be made to anything that has already happened (much as I would like to). Any future moment has yet to come and will be whatever it is because of events and forces far beyond me and my control. The only ability I have to affect my future depends solely on what I do at each present moment. But in our sophistry and our fear we also forget that, in the most real and practical sense, the future does not exist. We actually are alive and able to make choices only in a endless succession of present moments. Each of these "presents" is what it is because of all the simultaneous events and forces (including my own choices, perceptions and actions) in that moment and in the previous moments. Since we cannot affect the past moments, the only thing left is to make a different choice in the present.

Thus, to be able to make a different choice in any present moment (indeed any choice at all) means that you must be conscious enough to remember to do so.

However, once you realize that in most of your present moments you are not actually "present" — that is, there is no "you" there conscious enough to make any kind of choice (you are simply immersed like a baby in the womb in the content of what you are doing); and once you understand that when you are brought to awareness by a symptom (which is in effect a messenger telling you that the way you have been going about things is unconstructive and that your present experience is what it feels like to be operating that way) you tend to react against the messenger that woke you up rather than get the message; and once you admit that the way you react is to shoot the messenger as soon as it shows up so you can get right back to your narrowed and "unconscious" involvement in the details of your life; once you really get a sense of the totality of this pattern and its relentless mechanical cycling and recycling of your life, you will see that *the first step must be to have command of what will allow us to be more conscious in these on-going present moments so that we can have the possibility of making a choice.* Without this we are as bound and trapped as ever and will forever rebound from one manifestation of the habit to another, never being any the wiser.

So, what allows us to become more conscious? Let us sneak up on how consciousness works from its blind side, from how habitual ways of operating restrict and narrow consciousness.

It is my habit to feel that I cannot accomplish the job at hand unless I "concentrate", which for most of us means that we narrow our attention to cut out "distractions". Distractions are, of course, just other parts of this simultaneous moment. I do not realize that the reason I am distracted by what is actually happening around me is already because of my habit, but it does not take much to see how I collude with the habit in eliminating these "outside" aspects of my awareness. In fact, I *am* the habit.§§§ Daily, I am practicing and improving my ability to maintain a narrowed form of consciousness, and as Barbara Conable is fond of saying, "*practice does not make perfect — practice makes permanent*".

Because of this relative unconsciousness, when the habit cycles to the moment of the symptoms, I tend to respond as if there were no moments before, only moments ahead in which I can do something about them. The negative aspect of the feeling so dominates me that I am aware of almost nothing but the symptom and what I can do to get rid of it. These feelings then form another level of distraction, especially when they become chronic, which I work hard to ignore by further trying to narrow my awareness.

The same is true in the last part of the circle — that "end" I try to gain so all will be OK. As I flee from the symptom I am happy to aim all my attention to that projected future moment. As a consequence of that habitual striving and trying for the preconceived "good" solution (which has never really worked for me before), I am oblivious to all else that is in the moment — all the other unknown opportunities, all the possibilities inherent in my system just clambering to express themselves. In other words, every part of the circle fosters narrowedness and favours the known. This is why I end up being so "unconscious". But I am not really "unconscious". Rather, I am severely restricted in the extent of my awareness, which is to say, the extent of my being. This is what I feel as the symptom — my restricted narrowness of being.

Notice all the effort it takes to be so narrowed and unaware. Each step demands a massive input of energy to maintain. I *try hard to concentrate* on my job. There is a huge amount of *muscular work* we call tension that goes on while I am that narrowed. I expend

---

§§§ To paraphrase the comic-strip character Pogo, "We have looked the habit in the whites of the eyes and it is us"

124

an immense amount of *physical and emotional energy* trying to escape from the feeling of my narrowedness and I *strain mightily* to try to reach my ideal but impossible goal of being other than I am. I am deluded into thinking I am spending that energy to get somewhere, not realizing that it takes this much energy just to keep me out of the present, to keep me constricted, and to keep me in the tightly wound spiral of the habit. Day after day I am feeling the difficulty, the strain, the work and the cost, unaware that it is me supplying the energy that keeps it all running. I am supplying that energy sometimes willingly and sometimes unwittingly. One thing is for sure, there is no one else doing it for me and there is no one else doing it to me.

When we return to our question of what allows us to be more conscious, you can see that it is not a matter of how we can expand our consciousness, it is how can we stop constantly restricting and narrowing it. We are creatures that already have a wide-open and infinitely interconnected consciousness built into us. That is why your awareness and presentness expands to its innate openness in a few instants when you stop the end-gaining interference.

In this circular habit when is the only moment you are "naturally" brought to awareness? The moment of the very symptoms you love to hate. This is the moment when your wonderful highly-evolved-over-millions-of-years system has sent you a message to wake you up from your narrowedness. There is no possibility of changing anything or making any choices when we are "unconscious" in narrowedness. You may as well forget about that. The job is not to *try* to get consciousness into the areas where you are "unconscious", the job is to use constructively the moment when you already *have* consciousness and, fortunately for us, there will be plenty of times your system will wake you up if you are operating unconstructively.

If you can then meet the full force and reality of those moments and make the different choice to allow yourself to fully live these moments without reaction or judgement, you will find yourself more open and more conscious and present and hence more able to make these choices. Over time and with practice this will become your way of being — in other words, *a constructive circle* that reinforces itself.

When I can manage this simplest of simple choices, there is really no other choice needed from me. As I come more and more consistently into living in the present there are less and less "problems" that need any choices made about them. Most of what needs to happen in the moment is already occurring as a natural *response* when I am no longer *reacting.* These *responses* are not predetermined or preconceived by me, or my habit, or my culture. As I allow myself to open up I come into direct interaction with the events, situations and people around me. I am no longer working against the universe, I am inseparably part of everything that is. There is nowhere to get to any more.

I am home. ☺

Chapter Eight

A Resonance Model of Interaction
or, getting more from doing less

From a talk given in London, U.K., November 1993 at *The Centre for Training,*
an Alexander Technique Teacher Training program run by the author

I want to talk today about hands-on work, about a constructive way of thinking about what exactly is going on when you put your hands on someone and they begin to respond. Specifically, why does it seem that you have the greatest effect (sometimes surprisingly so) when you come up to someone and put your hands on them but you don't do anything at all to bring about change — so-called "non-doing hands"?

This is a hard one for new trainees, and sometimes even new teachers, to grasp because we often get very caught up in what seems like a common-sense idea that nothing will happen between one person and another unless there is some sort of actual physical "doing", some purposeful and skilled activity being carried out by the teacher that "causes" the changes in the student.

It is perfectly understandable that we'd have such an idea. After all, most of the things we do during the day do happen because we actually "do something". Certainly all the manual tasks we do with our hands are done by "cause and effect" manipulation: pouring the tea, changing a light bulb, extracting a sliver from your child's finger, even giving a massage to your lover... It's an idea we all have and which we bring into our first lessons in the Alexander work. The same idea is often reinforced in our early lessons when teachers put hands on us and we start to change. Of course, we'll tend to interpret it that our Alexander teachers have done something with their hands to bring about those marvellous effects on us.

Now, there definitely are some Alexander teachers who do direct manipulation with their hands, sometimes quite forcefully and sometimes to great effect. Other teachers do more subtle and softer manipulations, also sometimes to great effect. But let's be clear right off the bat, that neither of these approaches really deserve the term "non-doing hands". In fact, in many cases, that sort of direct hands-on "doing" is hard to distinguish from any number of other skilled manipulative body-work methods.

This "doing something" idea becomes a bit more subtle for those who still want to "do something" without "doing anything", at it were. There are variations, but it usually goes something like this: the teacher comes up to a pupil and puts their hands on and begins to pay attention to themselves to "get themselves going" into a good "Alexander state", and then also pays attention to the pupil as if they were pouring something out through their hands to the person. Whether what they are pouring through their hands is expressed as "direction" or "intention", it still is conceived of as a "cause and effect" activity. That is, the teacher with the skill and the idea of what the pupil needs is the cause and their hands are the means for that cause to get through and effect specific changes in the pupil.

With either of the above methods of hands-on — let's call them physically manipulative and intentionally manipulative — one thing is for sure. To go about it these ways the teacher must be able to see or sense in some way the specifics of the pupil's misuse (that held-up shoulder, that tightened back, that pulled-down front, that fixed hip, etc.), and also must have in mind an idea of the specifics of the goal — the desired better use of the

127

pupil (releasing the shoulder, lengthening the back, opening up the front, freeing the legs, etc.). Such a teacher will also tend to have a repertoire of some specific techniques of using the hands or specifics of thinking through a flow of direction that when brought to bear upon the pupil will, hopefully, bring about the desired changes. It is also part of this pedagogical approach that the teacher will have an eye out to note when those expected changes have happened, and will likely be interested in having the pupil recognize them too (in other words, training the pupil to have the same expectations of "good use").

If you've gone about things this way, you'll know how much work it is and how complex it gets, especially when the pupil is not exactly responding as you'd hoped. You'll also know how easy it is to get swept up in feelings of success when the hoped-for changes occur and swept away in a sense of frustration and failure when they don't.

But, we're not going to get into all that in this talk because it is all one form or another of a "doing" approach.

Instead, once you understand that "non-doing" hands can mean exactly that — doing absolutely nothing, trying nothing, and being after nothing — you also realize that it is the easiest thing in the world. After all, there is literally nothing to do! A challenging idea if ever there was one as you all have been discovering here on this training course.

So, what makes it so incredibly hard to do this incredibly easy thing? And why does it work so well when you do manage it? How can you get so much out of nothing? This is what I want to talk about today.

Let's start with a bit of background about the basic nature of the hands-on situation which is first and foremost about contact and connection between two people. Where you touch someone is not as important as the fact that you are touching them. In fact, the most important thing is not even that you are touching the pupil but that the two of you are close together for a time, sharing the same space, and being open to each other.

Second, the lesson situation is so much to do with intention — what both of you are sharing is an interest in whatever will happen in these moments of closeness and contact. That is, there is a mutually intentional space opened up to willingly experience and explore the moment as it is, and to allow possible changes to happen, hopefully, with no expectation about what they will be. This leaves the moment open to the unexpected and the new, to become whatever it will be.

Third, and in some ways even more important, is that the lesson situation (again hopefully) is about opening up the present moment of being, expanding the experience of what is and how that experience may be changing and what can be learned from it. This is different than what many are expecting lessons to be. Most pupils bring into the lesson their history of their chronic problems along with an expectation that they'll be able to get out of the moment they are in to a new and better moment.

Fourth, and finally, but not least, the hands-on session is a situation of response, not just between two beings, but between one highly trained being who (again hopefully) knows something, and more importantly who is living that knowledge, and one less experienced being who does not yet have that knowledge and skill but who wishes to learn it.

So today we are going to see how we can meet these aspects of the hands-on lesson and understand how people can actually respond to each other constructively without either one having to do anything at all to "make" or "get" that response.

128

In essence what happens is not that mysterious. In fact, as human beings we are built in such a way as to be sensitive and similar enough to each other that response between us is a natural phenomenon. In effect, there is an extremely powerful resonance that happens between human beings.

Partly, this happens simply because we're alive. It is in our nature as primates to be responsive to other members of our group. Humans are social creatures spending time together sharing the same space. We are drawn to be close and touch each other. We like the experience of connecting and bonding with each other.

Unlike the other primates we don't go around picking fleas out of each other's fur, but in a very real sense being close and touching each other is a human form of grooming. This goes for the Alexander Technique and, of course, many other activities too. But maybe I should not use the word "activities" here because it is not that we're doing some specific thing; it's simply because it's our nature.

In a sense, all creatures are built this way to some degree, but some creatures are much more that way than others. Tigers are solo animals and live for the most part all by themselves in the jungle staking out a territory that they don't allow other tigers to encroach upon (at least until mating time). Other creatures like ants are so locked into a larger group identity that, in effect, they don't really exist as individuals.

Primates are somewhere in between. We are definitely individuals and yet we need each other. We need the tribe; we need the group; and we need to be in connection because only then do we fully open up and find the full range of experience evoked by our responses to those around us. When we respond to others, we don't "do" it, we find "it happening to us".

However, it is not just in this larger sense that our responses have the characteristics of a form of resonance. I mean it in a much more physically grounded and particular sense too.

Resonance is a term in physics, chemistry and engineering that refers to certain properties of how one thing responds to, or is affected by, another. The phenomenon of resonance has a number of characteristics that define it, and I propose that these characteristics are also operative in us. As I describe them below, please keep in mind that I am not suggesting that resonance is an analogy that can be metaphorically applied to us. Rather I am making a case that we too exhibit the exact same resonant principles simply because we too are part of the same natural world.

Here are the characteristics of resonance in no particular order.

Similarity

First, is that any two entities, or structures since they do not have to be living, have the possibility of resonating with each other to the degree that they are similar to each other. Let's use an example to illustrate these characteristics.

You probably know what happens if you have two brass bells of similar shape hanging beside each other. When you clang one bell, the vibrations of the bell move the air around it in a series of compressed and rarefied sound waves which travel at the speed of sound over to the other bell. When this moving air "field", if I can call it that, encounters the second bell it is induced to move in sympathy so that it starts to ring as well.

Note that the second bell cannot prevent itself from ringing. It cannot stiffen up and say "No, I refuse to ring this time. I'm just not gonna do it... you can't make me..." It is pushed and pulled about by the moving sound waves and simply has to respond as the energy is absorbed and then "rebroadcast" by its own ringing. In other words, it begins to resonate with the first bell.

This happens because the two bells are of similar shape and made of the same or similar material. You could hang up a brass bell and then hang beside it a dead rat by its tail, but clanging the bell is not going to make the rat vibrate too. Even if you hang a cardboard bell of the same shape, clanging the brass bell is unlikely to set the cardboard bell ringing too.

Responsiveness

This brings us to the second characteristic which is that resonance is a function of the degree to which any two entities are able to respond to each other in a similar fashion.

This may seem obvious since almost by definition resonance occurs when one or both entities change to come into a sympathetic relation to each other. The bell can respond by changing from being still to vibrating. The dead rat cannot change at all except perhaps to stink a bit more tomorrow than today.

OK, you say, let's hang up a live responsive rat beside the bell. Now clang the bell, and I bet the live rat probably would respond, but not by ringing. Its "vibrating" and wriggling would not be in resonance to the bell, but in reaction to it.

In the same way, the cardboard bell does not respond like the brass bell. It is changeable, granted, in its own way — it will burn if set on fire — but it never will be able to respond with a similar pure tone as the metal bell. I'll have more to say about the relevance of this characteristic later, but it might be enough of a clue for now to say that more changeable entities will find it easier to come into resonance than more fixed or resistant (i.e. unresponsive) entities.

Proximity

In addition to similarity and responsiveness, there is also another factor that influences whether resonance will occur, and that is proximity. That is, the structures need to be close enough to affect each other. Imagine hanging two bells a kilometre apart; ringing one will have a negligible effect on the other one far away because the sound waves die out before they have gone a few hundred metres.

Now, let's look more specifically at these characteristics of resonance when two people come into a hands-on situation. First, similarity. As humans, of course, we are far more similar to each other than we are different. We may be different heights, weights, colours, and have different tastes, but even so, the similarities far outweigh the differences. We all have two eyes, a nose, two arms, two legs, more or less the same structural and neurological system, so the differences are miniscule compared to the similarities of being. We are far more similar to each other than we are to a monkey, to a rat (dead or alive), to chunk of broccoli, or to a rock, even one shaped like a human. So we definitely qualify in the similarity department.

In effect, what you are working with when you come up to someone, put your hands and allow response between you, are not the differences but the universal elements of all of us that humans are built to respond to (which is why it's a form of bonding). You're not working with the differences, you're working with the similarities. The differences become more and more irrelevant.

In fact, when you have an imposing way of working, that is, when you're trying to do something, trying to get to a predetermined goal, it tends to be the differences that are focussed on. The difference between you the teacher who has a job to do, and the pupil who has changes to make; the difference between the pupil as he or she is (with the problematic use) and the pupil as he or she should be and hopefully soon will be (with the better use)... However, when we come up to someone and impose nothing, be after nothing, not know what will happen, and allow both yourself and the pupil to be just who you are without judgment or pressure to change, then it will be those universal aspects of ourselves that are presented and that will be responded to.

Which brings us to the second characteristic: responsiveness. We are certainly responsive creatures. We can change in all kinds of ways. In an unconstructive direction we can sink into and be caught in any number of different forms of habits and poor use with their various internal separations and conflicts. And in the other direction we can come out of those habits and change toward that more universally recognisable state of better functioning with its unity and harmony. So we are definitely changeable creatures.

While speaking about responsiveness, let's be clear here that we have a potential responsiveness. Any one person may be open to change or quite fixed and resistant to change. A good part of your Alexander lessons and training has been aimed toward helping you become not only more changeable, but more open to change, that is willing to allow changes to come in and affect you even if they feel strange, or new, or disturbing.

On the other hand, you the pupil who first showed up lessons was likely far more fixed and held. Many pupils are not just set in their physical functioning, but also set in their mental and belief system habits. This makes them resistant to change as they hang onto the way they see things and the certainty they think their "reality" gives them. They think they have come to lessons in order to change, but their idea of the kind of change they want is to get rid of the problems without actually changing anything else.

As for proximity, we automatically get that because we've come up close to the other person and actually touched them.

So, we have the necessary conditions of similarity, responsiveness, and proximity to evoke resonance, but what happens once resonance does take place?

Harmony

I suppose the most obvious this is that when the two bells resonate with each other they come into harmony. They respond by entering into a mutual cooperative response — they go along with each other —rather than remaining separate (ignoring or resisting each other). You could say that to come into resonance is to come into a form of unity that brings the individual entities together into a shared and integrated wholeness.

Reinforcement

There is more to resonance than that, though. Not only does the first bell make the second bell begin to ring, but as they come into harmony they reinforce each other. The sympathetic ringing of the second bell sends out its own pressure sound waves that in turn reinforce the waves of the first bell. This not only makes it easier for both to keep ringing, but also intensifies and prolongs the sound of both, in effect, amplifying the harmonic resonance.

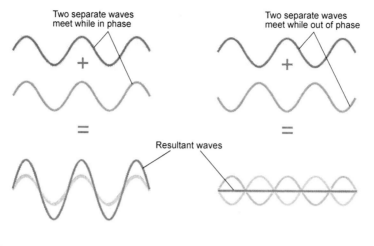

Two separate waves meet while in phase

Two separate waves meet while out of phase

Resultant waves

Two waves in phase *constructively interfere* to create an augmented or amplified wave

Two waves in opposite phase *destructively interfere* and cancel each other out

This is constructive interference, if you remember any of your high school physics. When the bells are in resonance, their sounds waves automatically meet each other in phase so that they add together becoming bigger and stronger than before.

Efficiency

What's great about this is that it doesn't take any energy to persuade the bells to make the precise changes so that they come into harmonic resonance. In fact, it takes less energy for them to resonate this way. To get the bells to not come into resonance while they are both ringing near each other is what would take extra energy. You'd have to try to get each bell to "ignore" the other bell in the face of its influence and ring without any regard for the other. With resonance you get more for less.

Affinity

Even better, it not only takes less energy to come into resonance, but it will tend to happen automatically. Similar, responsive entities close to each other have a tendency to come into resonance unless they are somehow prevented from doing so. In other words, they have an affinity or a draw toward harmony and resonance. It is easier to come into resonance when possible than to not do it… That is to say, resonance is a state that naturally and preferentially occurs because it is more efficient and "desirable" (if I can personify it) than non-resonance.

The phenomenon of resonance happens not only with inanimate objects like a bell but also with living creatures. As a human, you can probably relate to a lot of these characteristics already. Resonance also happens in man-made objects like electronic circuits where two similar enough circuits, like amplifiers, near each other will interact with each other to create beats and other rhythmic interferences, both constructive and destructive.

Spontaneous and Automatic

Another implication of this is that there is, in effect, a gradient that tends toward towards harmony and integration rather than disharmony and separation. I use the word

132

"gradient" in the sense of an incline or slope in which it is easier to go downhill than it is to go uphill. Naturally, if given a chance a ball will roll downhill. It does not take any energy. In fact, it releases energy. But it sure does take energy to push it back uphill...

So, similar, responsive entities close to each other will automatically and spontaneously tend to come into harmony, into resonance with each other, if there is any possibility of it. There is an in-built massive natural draw toward this happening. You would have to do something specifically and purposefully to stop it. It's just a property of nature that things will tend to come into harmony rather than expend the energy to maintain their separateness. It's simply what will happen with inanimate objects and living creatures to save energy and amplify our effect by coming into harmony and resonance. I'm sure that's why it feels good to us... because it is good for us!

Internal Harmony

Harmony not only happens between two creatures, but also internally in the degree to which you are functioning as an integrated whole as opposed to a collection of conflicted parts. "Integration" or "wholeness" are names we are likely to call this state of inner harmony where everything is working together and nothing sticks out. I'm sure you've experienced it. Every time you catch yourself in some sort of negative state: struggle, effort, tension, anxiety, etc., ask yourself: are you feeling whole or in parts? Are you feeling in harmony or in conflict? Are you expanded and open or narrowed and separate?

Isn't it true that every time things are working well, you can feel that you are more unified and whole in yourself? There is an ease with little or no effort and nothing to do. Everything is happening by itself, and you experience expansion and openness. In fact, not only are you not separated in yourself, but you are also not separate from the world around you. In other words, I'm not speaking theoretically here. You have had and will have many more experiences of this kind.

The Integration Gradient

If you have two systems which are similar and responsive there is a gradient or draw for them to "want" to come into resonance with each other. All fine and good, but in order to do that, at least one of them needs to change in some way to come into the harmony of resonance. Which one will make the change and why?

The answer lies in several important attributes of the gradient drawing us towards resonance.

One of these is that, simply put, if one of the systems is in a state of less integration (more separation) within itself, there will be a tendency for that less integrated system to be the one which changes. It will change by spontaneously reorganizing into a higher state of integration within itself in order to come into harmony with the more integrated system.

Obviously this "integration-separation gradient" can only occur in systems that have sufficient complexity which can be organized in either more integrated or less integrated states. A bell doesn't have this sort of inner complexity. It is cast as a single piece and can't be in more integration or less integration within itself. It can only ring or not ring. We, on the other hand, being made up of many parts and having many interconnected functions can be in a wide range of internal states, from un-integrated

133

The Integration Gradient: *Normally, the teacher will be more whole and integrated than the pupil. This means that the teacher's system is a big draw to induce a reorganization of the pupil's system towards more integration so that the pupil comes into resonance with the teacher. The resulting resonance has the maximum harmony (wholeness) and minimum energy cost (ease).*

states of conflicted parts within ourselves all the way to a state of harmonious unity within ourselves.

There is a reason why it is the less integrated system which will tend to come into a more integrated internal organization so as to come into harmony with the more integrated system. It happens that way because this results in a resonance which takes less energy and has more overall harmony. In other words, it is an easier state to come to and an easier state to maintain.

The alternative would be for the more integrated system to become less integrated (i.e. more disorganized) in itself in order to match the less integrated state of the other system. This certainly can happen as we will see in a moment, and it does result in a resonance which is more harmonious and uses less energy than if both systems remained separate and didn't change at all. The two systems are integrated with each other even if they are both at a relatively low level of integration within themselves. However, this option is not nearly as harmonious (efficient) as the first example where the less integrated system reorganizes to match the more integrated system, both of them now matching each other and at a high level of integration within themselves.

If you think about the implications of this for our hands-on work, you'll see that this is the gradient which automatically works in your favour — as long as you can remain integrated in yourself.

Let me use a simple diagram here of two individuals one of whom is you the teacher with hands on the other who is the pupil. The teacher (you) and the pupil are both complex but similar systems. However, one of the important differences between you is that, normally, as a result of the embodied knowledge you have gained in your training, you will be in a

134

more integrated (i.e. more whole and at ease) state within yourself than the pupil — maybe not all the time, but hopefully, more consistently than not. The pupil, new to the work, will likely be in a less integrated state (i.e. more in parts, with internal conflicts, tensions, effort and struggle).

This means that the integration gradient works in your favour because the pupil's system will tend to respond to your higher degree of integration by reorganizing towards more internal integration so as to come into resonance with your highly integrated system. There will be a big draw towards this just because you are close to each other and that is the way the gradient works, always seeking resonance where there is maximum harmony (more wholeness and togetherness) and minimum energy cost (ease and freedom).

The Fixedness Gradient

But wait. There is another difference between you and the pupil. You have spent years training as an Alexander teacher and you have changed many times — towards a better, more whole and more free use of yourself, then back to your old habits, then back to freedom again. In the process, you have become consistently more free (less fixed) and more open to change, as well as more able to allow yourself to respond in the moment than you were before. The pupil, however, will probably be much more fixed and narrowed (less free and open to change) than you are.

Unlike the first one, this particular gradient at first glance does not automatically work in your favour. In fact, it can work against you. On the other hand, it really is an essential

The Fixedness Gradient: *The teacher will likely be more free and open to change than the pupil. Since resonance will happen if at all possible, there will be a big draw for the changeable teacher's system to reorganize to match the pupil's more fixed system. The resulting resonance has a "harmony" between the two even if both are not very integrated internally.*

135

part of the hands-on situation, and there is a way to turn it to the benefit of the lesson. To understand why, we maybe need to look a bit more closely.

The pupil's habitual ways of being do not change much, if at all, from day to day. These habits are present in every moment and in every activity (the universal constant, remember). This is not just the pupil's stereotyped and fixed ways of holding and moving, but also their fixed ideas, fixed beliefs, fixed misconceptions, and fixed sensory interpretations. Because all of this is so "normal" and constant for the pupil, it is strongly held and strongly believed in. So much so that the pupil, for the most part, believes that it is "reality". And, if it is reality, then it just is. There is no other alternative. "I am heavy, it does take effort to stand up. Everybody knows that. What do you mean that I can stand up from the chair without effort! That's impossible. After all, I have to lift my weight up against gravity, don't I?"

For the pupil it is impossible that anything else could happen simply because it hasn't happened for such a long time. Or to put it in a slightly different way, there's nothing so fixed as a "reality" that has no idea there is anything outside of itself.

So, how does this openness-fixed gradient work? The moment you, with your more open and changeable system, come into contact with the pupil, your more changeable system is going to experience a massive draw to change and become organized like their more fixed system. This will happen simply because you are the more changeable one whereas they are fixed and cannot change. As long as the pupil remains fixed, if harmony and resonance is to take place between you both it will be because you are the one to change to come into resonance with the pupil.. In other words, when you are together there is a big draw for you to become as fixed as they are. It will take less energy for that to occur than for you to remain in your more integrated, open and changeable state beside them while they are more fixed.

Opposing Gradients: *The teacher's wholeness invites the pupil to reorganize at a higher level of integration. At the same time, the pupil's fixedness tends to make the teacher's changeability take on characteristics of the pupil's fixity and trying. The pupil may benefit somewhat, but the teacher will feel strain and hard work and ends up losing the very advantage the pupil initially responded to.*

136

You know this because I'm sure you have experienced it. When you come up into hands-on contact with a pupil and do not keep consciously choosing in an ongoing way (that is, spending the energy) to remain attentive and in the moment, expansive and open, you will find yourself becoming more and more narrowed to the pupil, less and less in the moment and probably more and more caught up in trying to change the pupil (end-gaining for another "better" moment), and eventually you'll find yourself more and more tense and held.

Interestingly, your pupil may experience some benefit in the moment, becoming a bit more free and a bit more integrated, as their system responds and converges part way towards yours. But, to the degree that your system gets organized towards theirs and becomes narrowed and fixated, you'll have lost the biggest asset you have — the very integration and openness that was the big draw for the pupil in the first place.

Do not mistake this change in the pupil for some success due to your hard work and trying. It is, in fact, the merest shadow of what could have been! And it has been your hard work and trying that has limited it all.

Well, you might say, doesn't this openness gradient contradict what I just said about the integration gradient? Indeed it does. They are opposites. The integration gradient is an automatic draw for the pupil's system to reorganize to a higher level of integration in order to come into resonance with you — it works for you. The fixedness gradient is also automatically drawing towards resonance by inducing whichever system is more able to change (you) to go through the reorganization to match the pupil's fixed state — it works against you.

On the other hand if you can keep yourself open and present with a certain level of integration within yourself in spite of the draw to be organized like the pupil, then you keep that strong inducement for the pupil's system to reorganize to your integration. This is the "work" you must do in a lesson to keep yourself "going", and if you do keep yourself going then this fixedness gradient does not work against you, but instead actually gives you something else essential to the teaching situation.

In the time you have been in this teacher training course you have had many opportunities to put your hands on others and you are getting better and better at doing this in a non-doing, non-trying, non-end-gaining sort of way. In the course of these times, I know you've experienced that curious situation where you have your hands on someone and you get an insight as to what is happening in their system. You might suddenly "sense" that their knees are locked or that they are holding their breath even though your hands are nowhere near those areas. How can this happen? You don't have x-ray eyes or telepathy… but you do have resonance.

Remember, the pupil's fixed state, fixed ideas, and fixed mode of being is a constant invitation — an attraction in the biggest sense of the word — that naturally draws you into being reorganized in a more fixed state yourself. Think what happens when the pupil's state affects you this way. In effect, their system gets into you, causing you to "mirror" them as resonance takes place[¶¶¶]. In other words, your system "picks up" what happens

---

[¶¶¶] My insight about resonance in hands-on work had been a mainstay of my teaching for many years before this talk in 1993. Around 1999 it was with some excitement that I first heard about so-called "mirror neurons". Discovered 1992, mirror neurons are neurons in the brain

in their system as it begins to respond, and that is how you get information about what is going on with the pupil.

It is not a direct "sensation" like your own tension would be. It's more like a hunch or an insight. But, as you've seen from your own experiences, it is an accurate perception about what is really happening in the pupil which you have verified by checking it out with them.

Interestingly, even though the pupil's system is trying to influence yours, to get you to respond to their habit, you do not feel it in yourself. You tend to feel it in the pupil.

If you are not able to manage to keep your own openness and integration going in the face of the pupil's less integrated and more fixed use, you will begin to resonate with their state and end up getting more narrowed and fixed yourself. This you will feel in yourself. But if you are able to keep leaving yourself open and whole, you won't be feeling, "Gosh, they must be fixed in their knees because I'm getting fixed in my knees". Instead, you will actually be sensing the other person's knees fix.

You feel this in the same way that when you take a stick by one end and move the other end around on a rough rock you can fell the

Flow of Information: As a teacher, your changeable and receptive system will be influenced by the state of your pupil. If you can remain open and integrated this influence won't get you caught up in the same fixedness as the pupil. Instead, it will give you information about what's happening with them.

roughness of the rock. You don't feel the vibrations of the stick on your fingertips and then infer that the rock must be rough. You feel the little bumps on the rock itself. You are not directly touching the rock and there are no sense receptors in the stick wired into your nervous system, yet it seems to you that you are feeling the texture of the actual rock.

Try it with a pencil or a pen now. Hold it by the pointy end so you don't make marks on anything, and then stroke the table top in front of you and ask yourself to notice whether the texture is smooth or rough. Notice when you perceive the texture your attention is down at the very end of the pen on the texture itself somehow feeling it "directly". It's quite a remarkable ability when you think about it. Your perceptual system, focussed by your question about the texture, undoubtedly is picking up vibrations from the table surface

---

which fire when an animal either makes an action itself or observes another animal making the same action. They allow us to mirror or copy another's actions. Mirror neurons have been found in primates and in birds and many neuroscientists believe that they are the foundation for imitative learning and the basis for empathy. I'm sure you can see the parallels to what I am describing here. They also seem to be implicated in *common coding theory* which states that there is a shared neural representation for both perception and action. Nevertheless, the exact mechanism by which these resonant responses take place is not of utmost importance for our purposes as teachers. What is important is that we recognize that the resonant response does happen and learn how to use it to our advantage in teaching.

138

transmitted through the pen to your fingertips. It then interprets these as reflecting the smoothness or roughness of the table. The amazing thing is that this perception is then projected back out so you feel it as coming, not from your fingertips, but the table itself. Your marvellous system is doing exactly what you asked it to do — giving you information about the table top. You don't have to figure anything out; your system does all the "calculating" and simply presents you with the answer in an easy-to-understand format.

In my opinion, this is the same thing that goes on in the resonance interaction between you and the pupil.

The pupil's fixed state is an insistent, almost stubborn, influence on your system to organize the same way. Provided you keep choosing to remain open and present, you'll remain receptive to this influence but not get sucked in. This allows your perceptual system to detect what's going on in the pupil's system. Of course, you are also highly trained to see things about the pupil's use and movements, and to directly feel all sorts of information. But this resonant "channel" is not from your eyes or your hands. It is using the whole of you as a receptor, as an "antenna of use" if I can put it that way.

However, as you begin to perceive what might be happening in the pupil, you face a choice. And that choice is what if anything to do with that information?

If you react to the information that the pupil is, say, tight in the shoulders by feeling that it is your job to get them free in the shoulders, you will likely begin to bring out some process from the Alexandery repertoire you have learned, the main purpose of which is to get the pupil to a state of more free shoulders. If you narrow your attention to part of the pupil (or yourself) you will lose your expansive openness. If you get caught into trying to change the present moment (tightness) into a better future moment (freedom) you will lose your presentness. If you direct parts of yourself in any way (neck, arms or hands), you will lose your wholeness.

Collectively these losses mean that your system is no longer whole, open, expansive and present in a way that exerts a big draw on the pupil to reorganize like you and come into resonance with you. Instead you will start to get reorganized like the pupil into effort, trying, end-gaining, narrowing, and ultimately tightness and frustration if you are not getting the results you want. In effect, you will have been organized into resonance with the student.

On the other hand, if you take in the information you have gained about the pupil and simply note it as showing you more about the person in front of you then you will literally be more in touch with the pupil as they are. The more you perceive the pupil, accept them, and allow them to be who they are no matter what sort of "use" they have, the more that your system is right there with them and the more that their system will be influenced by yours. As long as you maintain your wholeness and integration, remain open to whatever changes come along without predetermining what they should be, your system will present the hugest of influences, like a magnetic attraction, on the pupil's system to respond to you with a reorganization towards more wholeness and integration of their own.

But note, there is one big "if" here. Your free, open, whole, and integrated use of yourself is a huge influence on the pupil to reorganize like you... if, and only if, the pupil can allow themselves to become less fixed. As long as the pupil is holding on to their more fixed state (be that holding on physically and/or to fixed ideas), not much change toward a resonant response can possibly happen.

Of course, it takes energy to be fixed. For the pupil to hang onto a poor balance, force themselves into good posture, or move against their own stiffness takes muscular energy and shows up as effort, tiredness, and soreness. It takes mental and emotional energy to react when the world or other people are not the way you want them to be or when you are not the way you think you should be. This shows up as a variety of reactions like nervousness, judgment, frustration, impatience, conflict, anger, anxiety, doubt, shame, etc.

As I noted a ways back in this talk it also takes energy from you the teacher to stay present and open, that is, to meet the pupil's fixedness and expectations (and possibly your own expectations too) without reacting by trying to make things happen. This is not muscular or emotional energy, but you do have to spend energy to remain open to the moment as it is, to allow yourself and the pupil to be as you are and respond as you will, to choose not to react, to resist the temptation to try to get the pupil to where you think they should be. And, not least, to remain calm and encouraging, clear and communicative with your pupil even if they are confused about what they feel and what they should be doing to help, full of effort and trying, impatient to change and reaction when they do… This kind of choice and clarity from you does not come for free. You might call it the "cost" of teaching.

The purpose of spending the energy is to make sure that you're keeping your more-openness, more-integration in the face of the pupil's less-integration, more fixedness. Without that, your system will lose the big draw which has the biggest effect on the pupil. If you lose your more-changeableness by getting fixed into "trying", both of you will be putting more energy and effort into the lesson which is the opposite of what happens in harmony.

As long as you can keep that big draw of the "example" of your open and integrated system, then eventually your pupil will respond.

Let me illustrate this by toying with the analogy I just used of a magnetic attraction. Imagine that you have two large magnets, a "teacher magnet" and a "pupil magnet". When they come close to each other they exert a powerful force attracting them to move together. Normally each of them would move part of the way towards the other, but what happens when one is free to move and the other isn't? Well, obviously, the one which can move will move towards the one which cannot move.

To stretch the analogy even more, we want the less-integrated pupil magnet to move towards the more-integrated teacher magnet. Unfortunately, the pupil magnet is fixed and cannot easily move which is why there is a tendency for the more-moveable teacher magnet to be the one which does the moving, ending up stuck to the pupil with effort, tension, etc. However, if the teacher magnet can maintain its state of integration and grounding it cannot get pulled towards the pupil magnet so its pull will be constantly drawing the pupil toward it. But the pupil magnet cannot move towards the teacher magnet as long as it is fixed. It's an impasse unless something changes.

Perhaps while I have been gabbing around and around a few times about the way this works, you have already seen what needs to happen…

Somehow the pupil needs to come out of their fixedness. And this is where your next job as a teacher comes in. You need to use the information you gained as their fixed system influenced yours to help them become unfixed and be willing to change.

For pupils to respond they must be willing to try new things. Not just sort-of safe new things, but things that really feel weird and wrong and different… They need to be willing to entertain new ideas. Not just new ideas "intellectually", but ones with real repercussions, ones that contradict their cherished beliefs, ones that seem scary. They also need to be willing to feel themselves differently: to feel wrong, or bent or taller, or more vulnerable or… not themselves.

You will have some delightful pupils who are already game for all this but with other pupils it will be your job to inspire them, to encourage them, to explain things to them… to let you show them something truly new, to become less fixed, not only in the bodies and their ideas but also in their expectations..

Some of this you can accomplish by physically moving them to break up muscular and postural holding. Some of this you will need to use words to explain things or demonstrations to show things to break up fixed ideas or misconceptions. The main object with all of this is to get the pupil out of their fixedness and open to new responses. Once they're less fixed, and provided that you have been able to keep the big draw of your own wholeness and integration, resonance is automatically going to take place.

You can't do the integration for the pupil. It is not your job or your responsibility. It isn't even within your power. You don't "give them" the experience; you just allow yourself to be a good example, and help them open up so that integration can naturally occur. And integration will occur, sooner or later.

It's important at this point to stress that when this natural resonance happens, there are going to be two different kinds of integration occurring. One is that the pupil will be coming into a harmony of their own being, an integration within themselves. This is due to their own nature and has nothing whatsoever to do with you.

As you help them give up their physical fixedness and more fixed ideas, their system will be free to come into an integrated wholeness of the moment which is simply an expression of their own nature. This is their own version of the integration and wholeness that you have next to them. This change is not in response to you, but mostly in response to their own nature. It will be as much or as little as will possible for them at that moment.

In addition to that, of course, the opening up pupil is going to be in response to your system too. In effect, this is where the similarity of their integrated wholeness and your integrated wholeness begin to interact. You both begin to come into harmony and both affect each other in a different and more positive way than before — a way which begins to make it easier for both of you.

Moments before you the teacher had to spend some energy remaining open to your own integration, and at the same time spending some energy communicating with them to consider new ideas or demonstrating new possibilities. Once they begin to become less fixed and begin to come into the harmony of their own nature, and begin to interact with your nature, their harmony resonates with yours making it easier for you to keep yours because you don't have to keep it going against their fixedness.

Your integration is now harmonizing with their integration which is responding by becoming more integrated, which helps you let go of some of the energy to stay more integrated in the face of their fixedness. You get to be a little more easy and open which they respond to, which you respond to, which they respond to… and the whole thing becomes spontaneously so much more free and easy than it could be before.

They get to give up the effort of keeping their fixedness, and the cost of struggling with all the results of that fixedness — the tension, pain, anxiety, and the other stuff that burdens them with further struggle, treatments, searching, and so on.

141

In fact, it all becomes so easy that the only cost left is the cost of the lesson... (the class laughs).

You laugh, but I am not being facetious here. Harmony is such a desirable thing that nature will find ways to go towards it because it's simply built to flow that way. It takes energy to keep it out. So what the pupil, in effect, is really paying you for in a lesson is your knowledge of all this, your experience of how to help them learn it, and your willingness to spend this extra energy in the face of their habits that you wouldn't have to do if they weren't there.

Of course, you're willing to do that, and you're happy to help them learn how to incorporate it in their own lives so that they don't have to spend that energy all the time either. As you help them learn, they'll go out there into their world and be with people in a less strained and more present way, which has a positive effect on those people, and so it spreads. like the domino effect.

Over time it all radiates exponentially, so that more and more of us do not have to waste so much energy on our own conflicts or our separation from each other.

We'll be free to spend it on much better things...

Appendix

Ordering more copies of this book

Direct from the publisher — *Looking at Ourselves* is available directly from the publisher by mail order as a printed book (in both black and white and colour versions), or as an e-book which can be downloaded.
> **www.store.learningmethods.com#looking**

From other booksellers — The printed edition (black and white version only) and e-book editions should also be available at, or can be ordered from, major booksellers both online and in stores.

Quantity orders for Alexander Technique teachers:
Special discounts are available for quantity orders placed by certified Alexander Technique teachers or by bookstores ordering for Alexander Technique classes:
> **www.store.learningmethods.com#lookingat**

Library orders: Contact us at: **sales@learningmethods.com**

Other LearningMethods Publications:
You can find many other printed and audio-video publications in the LearningMethods Online Store at:
> **www.store.learningmethods.com/**

Bibliography and Further Reading

The Body Moveable
> by David Gorman, LearningMethods Publications 2002

Control of Head Movement
> by Peterson and Richmond, Oxford University Press 1988

Gray's Anatomy
> 36th edition or later, Churchill Livingstone

Man the Tottering Biped
> by Phillip Tobias, University of New South Wales 1982

Neurophysiology of Postural Mechanisms
> by T..D..M. Roberts, Butterworths 1978

Proprioception, Posture & Emotion
> editor, Dr. David Garlick, University of New South Wales 1982

The Lost Sixth Sense (booklet, 47 pages)
> Dr. David Garlick, University of New South Wales 1990

The Senses Considered as Perceptual Systems
> by James Gibson, Allen & Unwin 1968

The Wisdom of the Body
> by W. B. Cannon, Norton 1939

Integrative Action of the Human Nervous System
> by Sir Charles Sherrington, Yale Press 1947

Man on His Nature
> by Sir Charles Sherrington, Cambridge University Press 1951

The Spinal Engine
> by Serge Gracovetsky, Springer Verlag 1987

Physiology of the Joints
> (3 volumes) by Kapandji, Churchill Livinstone

Man and the Vertebrates
> (2 volumes) by A.S. Romer, Penguin 1968

The Alexander Review, 1986-1989
> editor, David Alexander, (no longer published)

Direction Journal
> editor, Paul Cook (web: www.directionjournal.com)

About the Author

David Gorman developed the LearningMethods work out of over 30 years of research and teaching experiences. He has a background in both science and as an artist and has always had a fascination with exploring human structure and function. In the mid-1970s he spent many nights dissecting and drawing in the human anatomy lab.

David has been studying and using the Alexander Technique since 1972 and has been teaching since the late 1970's becoming well-known worldwide for his innovations to the work and notorious for challenging the orthodoxy of the profession. He has been invited to teach all over the world in universities, conservatories and training colleges, at conferences and symposia, and with performance groups and health and medical professionals.

In 1980 he published an illustrated 600-page work, *The Body Moveable* and in 1997 the first edition of this collection of articles and essays, *Looking at Ourselves*. David has also been deeply involved in the professional organization of the Alexander Technique. He was one of the founding members of CanSTAT (Canadian Society of Teachers of the Alexander Technique), chairman of the founding committee of NASTAT (now AmSAT, the American Society for the Alexander Technique), the architect of the Affiliated Societies mechanism for Alexander Technique professional organizations to recognize each other, and the author of the original Sponsorship and Certification process of ATI (Alexander Technique International).

In 1982, his teaching was revolutionised by his discovery of a new model of human organisation with its profound implications of our in-built and natural tendency toward balance, ease and wholeness. He extended these insights into a new way of training teachers of the Alexander Technique and from 1988 to 1997 in London, U..K. he trained 45 teachers, assisted by Margaret Farrar and then Ann Penistan.

However, further explorations in his own and other training groups made it clear that a huge part of our problems lay not in the "body" but in our consciousness and way of seeing things -- our underlying belief systems and how we misinterpret daily experiences and then react to these misunderstandings. At this point it also became apparent that his discoveries and

146

the changed teaching methods they implied no longer could be squeezed in under the premises and pedagogy of the Alexander Technique.

Recognizing the need for a new and more effective approach to help people uncover and liberate themselves from these circular traps, David developed the LearningMethods work to teach people how to gain command of their exquisite in-built clarity of perception and powerful tools of intelligence so they can successfully navigate their lives.

Since the beginning of the work in 1997, David has completed the training of a growing number of LearningMethods Teachers, many of whom are now teaching the LM work in universities, colleges and conservatories. He continues to evolve the Apprenticeship Teacher Training Program having recently added online live video conferencing as part of the training. He gives workshops in Europe, North America and Asia as well as writing about the work and raising another young son.

Anyone interested in the work, upcoming courses, training as a teacher, etc. should visit the LearningMethods website for more information and contact details (www.learningmethods.com), or e-mail David directly at: david@learningmethods.com

About the LearningMethods work

We are the most amazing learning creatures on the planet. We have a wonderful in-built intelligence, an exquisitely sensitive awareness and a wealth of experience. Yet we still suffer from so many chronic and completely unnecessary problems.

Why?

Because only a few people have moved into their full heritage as humans and learned how to bring their intelligence to bear on their experience to understand and solve their problems.

How do I know this?

I see it every day in my teaching. Once people learn the simple yet incredibly powerful tools of how to unravel what's happening to them they can change far more easily and quickly than they dreamed possible.

How can this happen?

Because everything you need is already within you — the only thing missing is learning how how to use it.

LearningMethods teaches you how to use your intelligence to explore and understand your daily experience, to make sense out of your issues and problems, and how to change them. In other words, LearningMethods gives you the set of tools to navigate your life successfully -- to find your way back to your natural wholeness and presentness; to be free of chronic tension and effort; to be more effective and stress-free at work and to be more at ease and in harmony with others — in fact, the tools for solving any problem that's solvable.

It is not a far-off dream -- it is in your very human nature to be able to learn from experience and liberate yourself from problems. It is your birthright to gain wisdom about life and to feel at home in the universe. And, there is no time like the present to start !

David Gorman
founder of LearningMethods

Find out more about LearningMethods

Studying and upcoming workshops, finding teachers, reading on-line articles, publications and online store, teacher training and professional development:

LearningMethods homepage — www.learningmethods.com

What is LearningMethods? How it can help you, what you'll learn — www.learningmethods.com/lm-engl.htm

The principles of the work How does it operate? What is it based on? — www.learningmethods.com/question.htm

How do I study it? How LM is taught and the various ways to study it — www.learningmethods.com/lmstudy.htm

Upcoming Workshops Calendar of workshops and upcoming events — www.learningmethods.com/calendar.htm

Finding a Teacher List of LM teachers and apprentice-teachers — www.learning-methods.com/lmteach.htm

Quotes and Testimonials How LM has helped people and what they have said — www.learningmethods.com/lmquotes.htm

Online Library of Articles Exploring and solving various problems — www.learningmethods.com/lmarticle.htm

Online Store Buy printed publications, e-books, audio and video recordings — www.store.learningmethods.com/

Professional Development Online learning and professional development for teachers, practitioners or performers — www.training.learningmethods.com/

Training as an LM Teacher What's involved in becoming a LearningMethods or an Anatomy of Wholeness Teacher — www.training.learningmethods.com/

More Information on the Alexander Technique

Information about the Alexander Technique on the Internet

The Complete Guide to the Alexander Technique is a good general Alexander Technique Internet resource with many resources on the Alexander Technique — articles, audio and video recordings, podcasts, interviews and more: **www.alexandertechnique.com**

Alexander Technique Professional Organizations

Here is a listing of the current Alexander Technique professional organizations beginning with ATI, the international professional body with branches in 12 countries and members world-wide. All these bodies have a mechanism for assessing the competence of teachers and for granting Teaching Certificates and/or Teaching Membership. All have a Code of Ethics to which all Teaching Members subscribe.

Alexander Technique International (ATI)
Web: www.ati-net.com   *e-mail:* usa@ati-net.com
contact: www.ati-net.com/contact.php

Many countries have their own national Alexander Technique professional organizations. As of the publication of this book these are listed below. Details for each of these groups can be found at: www.stat.org.uk/pages/affiliated.htm (this link is to the web site of the UK-based STAT).

Australia — Australian Society of Teachers of the Alexander Technique   (AusTAT)

Austria — Gesellschaft für F.M Alexander Technik Österreich   (GATOE)

Belgium — Belgian Association of teachers of the Alexander Technique (AEFMAT)

Brazil — Associacao Brasileira da Tecnica Alexander (ABTA)

Canada — Canadian Society of Teachers of the Alexander Technique   (CanSTAT)

Denmark — Dansk Forening af Lærere i Alexanderteknik   (DFLAT)

Finland — Suomen Alexander-tekniikan Opettajat (FinSTAT)

France — L'Association Française des Professeurs de la Technique F. M. Alexander (AFPTA)

Germany — Alexander-Technik-Verband Deutschland   (ATVD)

Israel — Israeli Society of Teachers of the Alexander Technique   (ISTAT)

Netherlands — Netherlands Society of Teachers of the Alexander Technique   (NeVLAT)

New Zealand — Alexander Technique Teachers" Society of New Zealand   (ATTSNZ)

South Africa — South African Society of Teachers of the Alexander Technique   (SASTAT)

Spain — Asociación de Profesores de Técnica Alexander en España   (APTAE)

Switzerland — Schweizerischer Verband der Lehrer der F.M. Alexander Technique   (SVLAT)

United Kingdom — Society of Teachers of the Alexander Technique   (STAT)

United States — American Society of Teachers of the Alexander Technique (AmSAT)

LearningMethods Publications

For a complete list of LearningMethods Publications:

    – books & booklets   – downloadable e-books,
    – audio recordings   – video recordings

visit the LearningMethods Online Store:

www.store.learningmethods.com

– descriptions of the publications   – details of costs and shipping
– secure ordering by credit card   – instant downloading of e-books

Here are a few examples of what is on offer:

––––––––––––––––––––

THE BODY MOVEABLE — A detailed study of the structure, function and dynamics of the human musculoskeletal system with thousands of beautiful illustrations and clear hand-written text. See sample pages at link below.

600 pages, 1 volume gold-embossed hardcover
(large-format size: 9" x 12" / 228 x 305 mm)

A new edition will be out soon, likely in several versions: a one-volume hardcover version with black & white inner pages (like the current 5th edition), and two different three-volume softcover versions (one with black & white inner pages, the other with colour inner pages).

Full details, sample pages and ordering at **www.bodymoveable.com**

––––––––––––––––––––

LOOKING AT OURSELVES – Exploring and Evolving in the Alexander Technique

2nd print edition, 176 pages, paperback (in colour or B&W)
(6.69" x 9.61" / 170 x 244 mm)
www.store.learningmethods.com#looking
E-book: 174 pages (for PDF and popular e-book readers)
www.learningmethods.com/ebooks/gorman-looking.htm

IN OUR OWN IMAGE – The elastic suspension system.

This series of 8 articles is part of the *Looking At Ourselves* book above, and has been extracted to its own e-book.

E-book: 250 pages (for PDF and popular e-book readers)

www.learningmethods.com/ebooks/gorman-inourimage.htm

STANDING ON TOP OF THE WORLD – Experiencing the Fundamental Freedom of Support and Balance

This graduated series of practical and experiential follow-along demonstrations was developed over the last 15 years by David Gorman to teach the principles of

support and balance and how to live them in everything you do.

Audio, 29+ hours – 1 CD-ROM, with files in MP3 format

www.store.learningmethods.com#cdsupport

LIBERATING PERFORMERS – Eliminating Performance Anxiety (nervousness, self-judgment, doubt and fear)

A recording of the "Liberating Performers" series of workshops helping performers to see through the misconceptions that generate fear and nervousness, and how to get rid of it permanently.

Audio, 19+ hours – 1 CD-ROM with files in MP3 format

www.store.learningmethods.com#cdliberating

ON BALANCE AND SUPPORT – A video demonstration of some principles of balance and support, and hints for teachers about using them in classes.

Video, 1 hour – 1 DVD (region-free, PAL or NTSC format)

www.store.learningmethods.com#supportvideo

SOUNDS OF A PROVENCE SUMMER – Day & Night

Live recording of the quintessential sounds of the hot Provençal summer – the cigales (cicadas) by day and the crickets at night, plus a summer thunderstorm and soothing rainfall. Listen to a sample of the CD at the link below.

Audio, 76 minutes – 1 CD (playable in any CD player)

www.store.learningmethods.com#cdprovence

 **FREE E-BOOKS TO DOWNLOAD** — Browse a large collection free e-books versions of various articles. Download, read and send to friends…

> **www.store.learningmethods.com#ebooks**

AND COMING IN THE FUTURE…

UPCOMING RECORDINGS:

THE ANATOMY OF WHOLENESS – A workshop on our innate coordination of wholeness and ease and its implications for daily life, performance, healing.

Video, 15 hours – 5 DVDs in PAL or NTSC format

> **www.store.learningmethods.com#aow-video**

UPCOMING BOOKS:

 **REVEALING THE OBVIOUS**
– The Anatomy of Wholeness
A new book, still in process, on the coordination of wholeness and ease and its implications for you, the whole thinking responding being. Extracts will be available in e-book format soon.

> **www.learningmethods.com/revealing.htm**

STANDING ON TOP OF THE WORLD
– Experiencing the Freedom of Balance and Support
A book about a series of practical demonstrations that will put you in touch with your own system's signals about support and orientation so you can be supported, whole and easy, responsive and present in every aspect of your life.

> **www.store.learningmethods.com**

LIBERATING PERFORMERS
– Escaping the Trap of Nervousness, Self-judgment and Fear
An upcoming book on why we get trapped in performance and social anxieties, how these fears are based an interlocked series of misconceptions, and how understanding and using your in-built "value register" will liberate you permanently.

> **www.store.learningmethods.com**

152

Publishing History of the Articles in this Book

For the *Looking At Ourselves* **book as a whole:**

Published in printed paperback: ISBN: 1-9531990-02-0
   LearningMethods Publications, London (U.K.), October 1997
Published in PDF ebook format: ISBN: 978-1-897452-31-8
   LearningMethods Publications, Toronto, July 2007
First DNL eBook edition: ISBN 978-1-897452-17-2
   LearningMethods Publications, Toronto, July 2007
Second DNL eBook edition: ISBN 978-1-897452-53-0
   LearningMethods Publications, Toronto, December 2008
Second edition B&W printed paperback: ISBN: 978-1-897452-75-2
Second edition colour printed paperback: ISBN: 978-1-897452-76-9
Second edition PDF e-book: ISBN: 978-1-897452-80-6
   LearningMethods Publications, Toronto, January 2012

For the various individual articles:

Thinking About Thinking About Ourselves

First published on the LearningMethods web site,
   www.learningmethods.com, 1998
Published in printed A5-size booklet: ISBN: 1-903518-04-0
   LearningMethods Publications, London (UK), August 2000
Published in printed A5-size booklet: ISBN: 1-903518-05-9
in French translation
   LearningMethods Publications, London (UK), August 2000
Published in DNL webbook format: ISBN: 978-1-897452-32-5
   LearningMethods Publications, Toronto, June 2007
Published in PDF ebook format: ISBN: 978-1-897452-33-2
   LearningMethods Publications, Toronto, July 2007

In Our Own Image -- a series in 8 parts

First publication:
   Part 1: *The Alexander Review*, Vol.1, No.1, Jan. 1986
   Part 2: *The Alexander Review*, Vol.1, No.2, May 1986
   Part 3: *The Alexander Review*, Vol.1, No.3, Sep. 1986
   Part 4: *The Alexander Review*, Vol.2, No.1, Jan. 1987
   Part 5: *The Alexander Review*, Vol.2, No.2, May 1987
   Part 6: *The Alexander Review*, Vol.2, No.3, Sep. 1987
   Part 7: *The Alexander Review*, Vol.3, No.1, Jan. 1987
   Part 8: *The Alexander Review*, Vol.4, 1989
Series published in *Looking At Ourselves*, October 1997
   LearningMethods Publications, London, ISBN 0-9531990-0-2
Series published on the LearningMethods web site,
   www.learningmethods.com, 1999

Part 8 published as A5-size booklet, ISBN 1-903518-06-7
   LearningMethods Publications, London, August 2001
Parts 1-4 published as A5-size booklet, ISBN 1-903518-17-2
Parts 5-7 published as A5-size booklet, ISBN 1-903518-18-0
   LearningMethods Publications, London, October 2002
Series published in *ExchangE* (Journal of ATI):
   Part 1: ExchangE, Vol.3, No.1, Spring 1995
   Part 2: ExchangE, Vol.3, No.2, Fall 1995
   Part 3: ExchangE, Vol.4, No.1, Winter 1996
   Part 4: ExchangE, Vol.4, No.2, Summer 1996
   Part 5: ExchangE, Vol.5, No.1, Jan. 1997
   Part 6: ExchangE, Vol.5, No.2, July 1997
   Part 7: ExchangE, Vol.6, No.1, June 1998
Published in DNL webbook format: ISBN: 978-1-897452-16-5
   LearningMethods Publications, Toronto, June 2007

Teaching Reliable Sensory Appreciation
Published in DNL webbook format: ISBN: 978-1-897452-15-8
   LearningMethods Publications, Toronto, June 2007
Published in PDF ebook format: ISBN: 978-1-897452-18-9
   LearningMethods Publications, Toronto, July 2007

On Fitness
First published in *The Alexander Review*, Vol.4, 1989
Published on the LearningMethods web site
   www.learningmethods.com, 1999
Published in printed A5-size booklet: ISBN: 0-903518-07-5
   LearningMethods Publications, London (UK), August 2000
Published in DNL webbook format: ISBN: 978-1-897452-11-0
   LearningMethods Publications, Toronto, June 2007
Published in PDF ebook format: ISBN: 978-1-897452-20-2
   LearningMethods Publications, Toronto, July 2007

Overview — A World-wide Community
First published in *The Alexander Review*, Vol. 1, No. 1, Jan. 1986
Overview -- Emergence of a Profession
First published in *The Alexander Review*, Vol. 1, No. 3, Sept. 1986
Overview -- Recent Developments
First published in *The Alexander Review*, Vol. 2, No. 1, Jan. 1987
Overview -- Looking Outside Ourselves
First published in *The Alexander Review*, Vol. 3, No. 1, Jan. 1988
Experience and Experiments in the Alexander World
First published in *The Congress Papers* by Direction Journal, July 1992

Publishing History, continued…

Published on the LearningMethods web site,
www.learningmethods.com, July 1992

The Rounder We Go, The Stucker We Get

First published under original title: *"The Rounder We Go, The Samer We Get"*
on the LearningMethods web site, www.learningmethods.com, 1999
Published under original title: *"The Rounder We Go, The Samer We Get"*
in printed A5-size booklet: ISBN: 1-9531990-2-9
LearningMethods Publications, London (UK), August 2000
Published in DNL webbook format: ISBN: 978-1-897452-15-8
LearningMethods Publications, Toronto, June 2007
Published in PDF ebook format: ISBN: 978-1-897452-18-9
LearningMethods Publications, Toronto, July 2007

A Resonance Model of Interaction

Not published before this edition

Index

A

B

inhibit 57, 68, 77, 79, 84, 119
inner ear 65, 66, 68
In Our Own Image i, ii, 15, 17, 21, 25, 29, 35, 42, 48, 55, 103, 153
instability 3, 6, 7, 26, 28, 29, 32, 33, 34, 35, 38, 39, 40, 43, 46, 48, 49, 50
integration 12, 79, 89, 90, 91, 108, 132, 133, 134, 135, 136, 137, 138, 139, 140, 141
intention 19, 32, 35, 39, 40, 44, 58, 66, 101, 104, 127, 128
intercostals 50, 51
internal oblique 51
internal register 89
intervertebral discs 21
isometric 91

J

jaw 51, 52, 53
joints, distortion 7, 22, 23, 25, 27, 28, 61
joints, moveable 21
Jones, Frank Pierce 58, 102

K

karate 89
kinesiology 15, 40
kinesthetic reality 5
Kirby, Ron 15
kneecap 34, 49

L

larynx 53
latissimus dorsi 51
LearningMethods ii, 56, 73, 144, 145, 146, 147, 148, 150, 153, 154, 155
left brain 11, 12, 13
lengthening device 19, 20, 22
lesser trochanter 34
levator costarum 50
levator scapulae 50
life support organ systems 20
ligaments 10, 18, 19, 21, 29, 34, 49, 63
lungs 20, 23, 53

M

manner of use 18
massage 96, 113, 127
mesentery 53
metabolism 24, 38, 56
motor nerves 32
moveable joints 21, 23, 25, 27, 50
Murray, Alex 95
muscle fibres 29, 30, 31, 32
muscles

abdominal 9, 24, 36, 51
muscle spindle 31, 32, 61, 102
musculoskeletal ii, 24, 52, 53, 55, 60

N

NASTAT iii, 99, 100, 101, 102
Nautilus 91
neuromuscular 88
Neurophysiology of Postural Mechanisms 15, 145
Newton, Sir Isaac 2
Nielsen, Michael 102

O

oesophagus 53
On Fitness 73, 154
oppositions 64
optimal environment 20, 23, 25
optimize 57
organ systems 20, 21, 23, 24, 25, 48, 52, 57, 59
osteopathy 96
otolith organs 65
Overview 94, 154

P

pectoralis major 50
pectoralis minor 50
pelvis 9, 18, 20, 21, 22, 24, 28, 33, 34, 36, 49, 51, 52, 53, 63
pharyngeal muscles 52
physiology ii, 33, 40, 55, 60, 74
Pogo 124
point of view i, 2, 3, 4, 5, 8, 9, 11, 12, 13, 17, 26, 35, 36, 37, 42, 45, 48, 55, 57, 59, 60, 61,
        62, 63, 64, 73, 77, 78, 80, 81, 84, 85, 89, 90, 92, 93, 101, 104, 115
poise 59
poor use 17, 89, 131
positional description 78, 79
posterior serratus inferior 51
posterior serratus superior 50
postural reflexes 59
posture 7, 10, 15, 22, 24, 26, 28, 35, 40, 46, 47, 59, 74, 75, 77, 78, 89, 140
  bad 24, 59
  good 22, 24, 28, 35, 59, 77, 140
potential energy 25, 26, 39, 40
pre-sprung elastic suspension system i, 15
pre-technique 89
primary control 19, 20, 55, 56
primates 25, 129, 138
Probable Reality Behind Structural Integration 15
proprioception 61, 66

psychological 112
psycho-physical unity 122
pubic bones 9, 21, 34

Q

quadratus lumborum 51
quadriceps 34, 49, 91

R

reactive cycle 112
reality ii, 1, 4, 5, 6, 9, 13, 47, 59, 60, 83, 85, 104, 117, 118, 120, 125, 131, 136
rectus abdominus 51
reflex 10, 11, 18, 32, 33, 35, 39, 40, 57, 58, 59, 66, 67, 68, 69
reliable sensory appreciation 47, 73, 154
resilience 22
respiration 24
respiratory 20, 23, 52, 53
reticulin 29
rhomboids 50
rhythm 24, 92
right brain 12, 13
Roberts, T.D.M. 15, 145
rowing 9, 88
running iii, 34, 39, 42, 45, 56, 88, 125
Running Without Fear 87

S

sacro-iliac joints 22
sacrum 21, 22, 34, 48, 49
scalene muscles 50
scapula 18
semicircular canals 65, 68
serratus anterior 50
shoulder blades 22, 45, 50, 64, 92
skeleton 4, 10, 27, 37, 40, 42, 43, 48, 50
slouching 4, 8, 23, 24, 39, 50
soleus 49
spacer 20, 40, 42, 43, 45, 46, 50
spinal column 21, 22
spinal cord 32
spine 7, 8, 15, 19, 20, 21, 22, 23, 24, 27, 33, 38, 46, 48, 49, 50, 51, 52, 53, 54, 60, 61, 64, 67, 75, 81
spring-loaded ii
STAT ii, iii, 1, 56, 99, 100, 101, 102, 149
sterno-mastoid 50, 51
sternum 22, 50, 51, 54
Stevens, Chris 102
stomach 24, 36, 45, 53

strengthening 9, 36, 37, 87
stretch-receptor 32
stretch reflex 18, 32, 33, 35, 39, 40
suboccipital 60
supra-hyoid 52
suspension system i, 15, 16, 40, 42, 44, 46, 47, 48, 50, 52, 54
symptom 112, 113, 114, 115, 116, 117, 118, 124
synovial 21, 22, 23, 27, 28, 61
synovial joints 21, 22, 23, 27, 28, 61

T

t'ai chi 89
tendons 10, 19, 21, 29, 66
tensegrity 16, 43, 44, 45, 52
tensor fasciae latae 49
The Body Moveable ii, 15, 55, 145, 146, 150
The Complete Book of Running 88
The Probable Reality Behind Structural Integration 15
The Tottering Biped 89
The Use of the Self 70
throat 20, 51, 52, 53
tibia 34
tibialis anterior 34
Tobias, Philip 89
toe extensor muscles 34
tongue 52
trachea 53
training course iii, 62, 102, 108, 128, 137
trapezius 50

U

unconstructive habit 112
unreliable sensory appreciation 80
uprightness 3, 8, 25, 26, 27, 31, 33, 34, 35, 39, 40, 44

V

value-system 89
vasti muscles 49
veins 53, 87
vertebra 20, 21, 22, 23, 24, 34, 49, 50, 52, 61, 64
vertebrates 23, 25, 57
vestibular apparatus 65, 70
vicious circle 9, 37, 59, 87, 112, 114, 119
virtuous circle 84
visually present 67
voice 18, 96
vulnerable 83, 141

W

Z

CPSIA information can be obtained at www.ICGtesting.com
Printed in the USA
BVOW061834150513

320818BV00005B/7/P